Lecture Notes in Computer Science   13253

More information about this series at https://link.springer.com/bookseries/558

John Gustafson · Vassil Dimitrov (Eds.)

Next Generation Arithmetic

Third International Conference, CoNGA 2022
Singapore, March 1–3, 2022
Revised Selected Papers

Springer

Editors
John Gustafson
School of Computing
National University of Singapore
Singapore, Singapore

Vassil Dimitrov
University of Calgary
Calgary, AB, Canada

ISSN 0302-9743 ISSN 1611-3349 (electronic)
Lecture Notes in Computer Science
ISBN 978-3-031-09778-2 ISBN 978-3-031-09779-9 (eBook)
https://doi.org/10.1007/978-3-031-09779-9

This Springer imprint is published by the registered company Springer Nature Switzerland AG
The registered company address is: Gewerbestrasse 11, 6330 Cham, Switzerland

Preface

As the shrinking of transistors (Moore's law) is hitting physical limits, those in the fields of high-performance computing (HPC) as well as those pursuing artificial intelligence (AI) are exploring other ways to perform more computing. This has led both groups to explore approaches to computer arithmetic that break from traditional fixed-point and floating-point representation.

As part of SCAsia 2022, the Conference on Next-Generation Arithmetic (CoNGA 2022) provided the premier forum for discussing the impact of novel number formats on

- application speed and accuracy,
- hardware costs,
- software-hardware codevelopment,
- algorithm choices, and
- tools and programming environments.

This was the third event in the CoNGA conference series. The submissions for the technical papers program went through a rigorous peer review process, undertaken by an international Program Committee. A set of eight papers were finally selected for inclusion in the proceedings. The accepted papers cover a range of topics including image processing, neural networks for machine learning and inference, encoding of multidimensional real number arrays, and improved decoding methods. The emerging posit format is considered in most of these papers, and is compared with proposed floating-point formats from Google, Nvidia, IBM, and others. After five years of effort, the Posit Standard was ratified by the Posit Working Group as the result of CoNGA 2022 meetings that resolved all remaining issues.

We would like to thank all authors for their submissions to this conference. Our sincere thanks go to all Program Committee members for providing high-quality and in-depth reviews of the submissions and selecting the papers for this year's program. We also thank the conference organizers for giving us the opportunity to hold CoNGA 2022 as a sub-conference of SCAsia 2022.

April 2022

John Gustafson
Vassil Dimitrov

Organization

General Chair

John L. Gustafson   National University of Singapore, Singapore

Vice Chair

Richard Murphy   Gem State Informatics Inc., USA

Technical Papers Chair

Vassil Dimitrov   University of Calgary and Lemurian Labs, Canada

Program Committee

Shin Yee Chung	NSCC, Singapore
Marco Cococcioni	University of Pisa, Italy
Himeshi De Silva	A*STAR, Singapore
Vassil Dimitrov	University of Calgary, Canada
Cerlane Leong	CSCS, Switzerland
Peter Lindstrom	Lawrence Livermore National Laboratory, USA
Richard Murphy	Gem State Informatics Inc., USA
Theodore Omtzigt	Stillwater Supercomputing, USA

Contents

Contents

On the Implementation of Edge Detection Algorithms with SORN Arithmetic

Moritz Bärthel[1]([✉])(iD), Nils Hülsmeier[1](iD), Jochen Rust[2](iD), and Steffen Paul[1](iD)

[1] Institute of Electrodynamics and Microelectronics (ITEM.me),
University of Bremen, Bremen, Germany
{baerthel,huelsmeier,steffen.paul}@me.uni-bremen.de
[2] DSI Aerospace Technologie GmbH, Bremen, Germany
jochen.rust@dsi-as.de

Abstract. Sets-Of-Real-Numbers (SORN) Arithmetic derives from the type-II unums and realizes a low-complexity and low-precision digital number format. The interval-based SORNs are especially well-suited for preprocessing large datasets or replacing particular parts of threshold-based algorithms, in order to achieve a significant reduction of runtime, complexity and/or power consumption for the respective circuit.

In this work, the advantages and challenges of SORN arithmetic are evaluated and discussed for a SORN-based edge detection algorithm for image processing. In particular, different SORN implementations of the Sobel Operator for edge filtering are presented, consisting of matrix convolution and a hypot function. The implemented designs are evaluated for different algorithmic and hardware performance measures. Comparisons to a reference Integer implementation show promising results towards a lower error w.r.t. ground truth solutions for the SORN implementation. Syntheses for FPGA and CMOS target platforms show a reduction of area utilization and power consumption of up to 68% and 80%, respectively.

Keywords: SORN · Unum · Computer arithmetic · Image processing

1 Introduction and Related Work

The universal number format unum, proposed by John Gustafson [12], presents a new approach for the computation with real numbers in digital hardware systems. To enhance and overcome traditional number formats, especially the IEEE standard for floating point arithmetic [16], the initial type-I unums were designed to utilize Interval Arithmetic (IA) instead of rounding in order to avoid the propagation of rounding errors. In addition, type-I unums exploit variable mantissa and exponent lengths for a reduced datapath and memory bandwidth. Evaluations and discussions on unum type-I hardware implementations can be found in [5,10] and [2].

Based on the initial unum format, with type-II unums and the corresponding Sets-Of-Real-Numbers (SORN) [11], as well as type-III unums (posits) [13], two

J. Gustafson and V. Dimitrov (Eds.): CoNGA 2022, LNCS 13253, pp. 1–13, 2022.
https://doi.org/10.1007/978-3-031-09779-9_1

further formats were derived. Whereas posits provide a less radical approach with constant bit lengths that can be used as a drop-in replacement for other floating point formats with compatibility to legacy systems, type-II unums and SORNs utilize the implicit IA concept created for type-I unums and radicalize this approach towards a very low precision format enabling low-complexity, -power and -latency implementations of arithmetic operations. A detailed introduction to SORN arithmetic is given in Sect. 2.

Due to the low-precision nature of SORNs, the format is not applicable to any application or algorithm. However, it can be shown that SORNs are especially well suited for preprocessing large systems of equations in order to reduce the amount of solutions for a certain optimization problem, such as in MIMO detection [4] or training of Machine Learning algorithms [14]. Another suitable application for the low-precision SORN arithmetic are threshold-based algorithms were a high accuracy result is not of major interest, as long as a sufficient threshold detection can be provided. In this work such a threshold-based algorithm for image processing is implemented and evaluated for SORN arithmetic. In particular, the Sobel Operator [18] used for edge detection in images is implemented as a full SORN and a hybrid Integer-SORN design and compared to an Integer reference design. Details on the Sobel Operator and the respective SORN implementations are given in Sec. 3. FPGA and CMOS synthesis results, as well as an algorithmic evaluation of the different Sobel implementations based on a reference image data set are provided in Sect. 4.

2   Type-II Unums and SORNs

One of the main concepts of type-I unums is implicit IA by means of an extra bit after the mantissa, which indicates the presence of an open interval whenever maximum precision is exceeded [12]. With this approach, rounding errors can be omitted at the expense of a certain imprecision, when an open interval is given as result of a computation instead of a single value. Type-II unums fully utilize this interval concept by reducing the representation of the real numbers to only a small set of exact values and open intervals.

2.1   Original Type-II Unums and SORNs

For the original type-II unum representation proposed in [11], the real numbers are represented by a set of n exact values called *lattice values* l_i, including zero $(l_0 = 0)$, one $(l_{(n-1)/2} = 1)$ and infinity $(l_{n-1} = \infty)$, and the open intervals in between. Every lattice value is included with a positive and negative sign. A basic set with $n = 3$ is given with the lattice values $l_i \in \{0, 1, \infty\}$:

$$\{\pm\infty \ (-\infty, -1) \ -1 \ (-1, 0) \ 0 \ (0, 1) \ 1 \ (1, \infty)\} \tag{1}$$

The representation can be extended by introducing further lattice values $l_i > 1$ and their reciprocals to the set. A general representation can be interpreted as depicted in Fig. 1a.

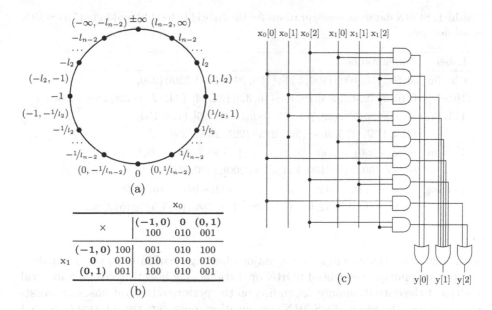

Fig. 1. (a) Representation of the reals with the original type-II unum format. (b) LUT for the multiplication of a simplified 3 bit SORN datatype. (c) Gate level structure for the 3 bit SORN multiplication LUT.

For the implementation of arithmetic operations, the so-called Sets-Of-Real-Numbers (SORN) binary representation is derived from the unum type-II set. The absence (0) and presence (1) of every lattice value and interval is indicated with a single bit, resulting in a SORN bitwidth $w_{\mathrm{sorn}} = 2^n$. Arithmetic operations with SORNs are carried out using pre-computed lookup tables (LUTs) which contain the result of every possible input combination for a given datatype configuration. Figure 1b shows the LUT for the multiplication of two SORNs using a simplified 3 bit datatype. Some SORN operations may result in union intervals, for example when two open intervals are added. In this case the result is represented by a pattern of consecutive bits:

$$100_{(-1,0)} + 001_{(0,1)} = 111_{(-1,1)} \tag{2}$$

The LUT structures for SORN operations can be implemented for hardware circuits using simple Boolean Logic which enables very fast computing with low-complexity. The corresponding gate level structure for the multiplication LUT in Fig. 1b is depicted in Fig. 1c.

2.2 Adaptions of the SORN Representation

Following the regular unum type-II-based structure for implementing SORNs maintains the unum compatibility and provides an error-free solution for processing arithmetic operations. However, the structure of the LUT-based arithmetics

Table 1. SORN datatype configurations for the hybrid (6 b–11 b) and full (15 b) SORN sobel designs.

Label	Configuration
6 b lin	$[0, 50]; (50, 100]; (100, 150]; (150, 200]; (200, 250]; (250, \infty]$
10 b log	$0; (0, 2]; (2, 4]; (4, 8]; (8, 16]; (16, 32]; (32, 64]; (64, 128]; (128, 256]; (256, \infty]$
11 b lin	$[0, 25]; (25, 50]; (50, 75]; (75, 100]; (100, 125]; (125, 150];$
	$(150, 175]; (175, 200]; (200, 225]; (225, 250]; (250, \infty]$
15 b lin	$[-\infty, -300); [-300, -250); \ \dots \ ; [-100, -50); [-50, 0); 0;$
	$(0, 50]; (50, 100]; (100, 150]; (150, 200]; (200, 250]; (250, 300]; (300, \infty]$
15 b log	$[-\infty, -512); [-512, -256); \ \dots \ ; [-32, -16); [-16, 0); 0;$
	$(0, 16]; (16, 32]; (32, 64]; (64, 128]; (128, 256]; (256, 512]; (512, \infty]$

with low bitwidths encompasses a major challenge within complex datapaths: computing multiple sequential SORN operations may lead to increasing interval widths at the output, mainly depending on the performed operations. In a worst-case scenario, the result of a SORN computation represents the interval $(-\infty, \infty)$ and does not contain any useful information. This can be counteracted with a higher resolution within the SORN representation. Evaluations in [3] showed that the exact values within a unum-type-II based SORN are barely ever addressed without their adjacent intervals. Consequently, moving away from a strict unum type-II based structure and adapting the SORN representation towards half-open intervals without exact values increases the information-per-bit within a SORN value and reduces the interval growth. Possible SORN representations following this concept are given in Table 1. The corresponding label indicates the number of elements in the Set-Of-Real-Numbers, which is also the number of bits in SORN representation. In addition, the label indicates whether the intervals within the set tile the real number line in a logarithmic or linear manner. In order to find a suitable datatype for a given application, the automatic SORN datapath generation tool from [17] provides an easy and fast way of prototyping SORN arithmetics for hardware circuits.

3   Edge Detection

In this work SORN arithmetic is applied to the Sobel Operator, an algorithm used for edge detection within image processing systems. Edges are regions in a digital image where distinct changes in color or brightness can be detected [1], in order to classify segments of the image, or to detect certain objects. Edge detection is used in various modern applications, such as fingerprint recognition [7], cloud classification via satellite images [8], or autonomous driving [6,20].

3.1   Sobel Operator

The Sobel Operator belongs to the family of first-order convolutional filters that compute the horizontal and vertical gradient of a grayscale image [18]. The Sobel method uses two 3×3 kernels, which are convolved with the grayscale image $\mathbf{A} \in \mathbb{N}^{N_x \times N_y}$ in order to determine the image gradients G_x and G_y in horizontal and vertical direction, respectively [19]:

$$G_x = \begin{bmatrix} 1 & 0 & -1 \\ 2 & 0 & -2 \\ 1 & 0 & -1 \end{bmatrix} * \mathbf{A}_{3 \times 3} \qquad G_y = \begin{bmatrix} 1 & 2 & 1 \\ 0 & 0 & 0 \\ -1 & -2 & -1 \end{bmatrix} * \mathbf{A}_{3 \times 3} \qquad (3)$$

After computing the image gradient

$$G = \sqrt{G_x^2 + G_y^2} \qquad (4)$$

a comparison to the pre-defined threshold T determines whether the current pixel is an edge. This process is performed for every single pixel of the image $\mathbf{A}$ and results in a binary image containing all detected edges.

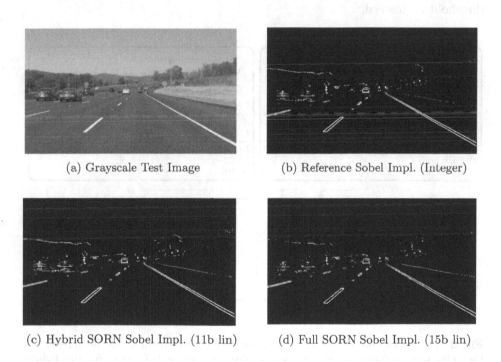

(a) Grayscale Test Image   (b) Reference Sobel Impl. (Integer)

(c) Hybrid SORN Sobel Impl. (11b lin)   (d) Full SORN Sobel Impl. (15b lin)

Fig. 2. (a) Grayscale highway test image [9] with Sobel edge detection results from (b) an Integer reference implementation with threshold $T = 250$, (c) a Hybrid-SORN 11 b implementation with threshold interval $T = (250, \infty]$, and (d) the negated result for a full-SORN 15 b implementation with threshold interval $T = (0, 50]$.

In Fig. 2 edge detection applied to a highway image is shown, which is used for road lane detection in driving assistant systems [6,9]. Figure 2a shows the grayscale test image and Fig. 2b the result of an edge detection using the Sobel method with integer arithmetic.

3.2 SORN Implementation

In this work, the Sobel method described in Eq. (3) and (4) is implemented with SORN arithmetic as a hybrid Integer-SORN and as a full SORN design, both for different SORN datatypes. Additionally, an Integer reference design is implemented in order to compare the SORN designs to a State-of-the-Art (SotA) architecture. The three designs are described in the following.

Integer Reference Design. The grayscale test image **A** contains pixels with values $A_{xy} \in \{0, \ldots, 255\}$ which can be implemented with Integer values of 8 b. The convolution described in Eq. (3) is implemented with conventional Integer additions and subtractions as shown in Fig. 3. For the calculation of the gradient G the square root is omitted and the result G^2 is compared to the squared threshold T^2 instead.

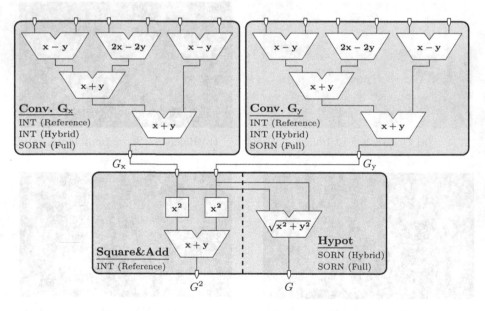

Fig. 3. Block diagram for the three different Sobel implementations: all integer for the reference implementation, integer convolution with SORN hypot for the hybrid approach, and SORN convolution and hypot for the full SORN approach.

Hybrid Integer-SORN Design. For the hybrid design, the convolutions are carried out with Integer operations, similar to the reference implementation. The horizontal and vertical gradients G_x and G_y are then converted to a SORN representation. Since they are squared in the following hypot operation, their absolutes are converted, and SORN datatypes without negative values can be used. The hybrid design is implemented for three different SORN representations with 6 b, 10 b and 11 b, all given in Table 1. The hypot operation is implemented as one single SORN operation, as depicted in Fig. 3 (conversion blocks from Integer to SORN between convolution and hypot are not shown). Since the result G is in SORN representation, the threshold T has to be chosen as one of the SORN intervals from the respective datatype. Figure 2c shows the edge result of the hybrid SORN implementation for the 11 b datatype and the threshold interval $T = (250, \infty]$, which corresponds to the Integer threshold $T = 250$ used for the reference implementation in Fig. 2b.

Full SORN Design. For the full SORN design, the Integer inputs from the test image $\mathbf{A}$ are converted to SORN representation before the convolutions and hypot function are carried out in SORN arithmetic, as shown in Fig. 3 (conversions not shown). Since for the convolution also subtraction is required, the full SORN design is implemented for two different 15 b datatypes with negative values, as shown in Table 1. In order to obtain a comparable edge result, for the full SORN approach thresholds near the zero-bit in SORN representation are selected and the result image is negated afterwards. Figure 2d shows the negated edge result for the full SORN implementation with the 15 b lin datatype and a threshold interval $T = (0, 50]$.

4 Evaluation

Figure 2 shows a test image and the edge detection results of the three different Sobel implementations described in Sect. 3.2. By visual comparison they seem to be quite similar, even though a few differences can be found, for example when comparing the detection of the cars on the road. For a comprehensive evaluation, however, a visual comparison of different edge results is not sufficient. Unfortunately, measuring the performance and comparing different edge detection methods or implementations is an open problem. In [15] various error and performances metrics are evaluated and compared, and the authors conclude that no convincing general-purpose solution exists. Since in this work no different methods, but only different implementations are to be compared, the most intuitive approach is a numerical comparison of the different edge results. Therefore the normalized absolute error nae between the SORN results and the Integer reference implementation can be defined as

$$nae = \frac{\sum_{x=1}^{N_x} \sum_{y=1}^{N_y} (E_{\text{int}}(x, y) \neq E_{\text{sorn}}(x, y))}{N_x N_y} \tag{5}$$

with the respective edge detection results E_{int} and E_{sorn} and the test image dimensions N_x and N_y. This metric basically counts the number of different pixels between the Integer and SORN edge images and normalizes the result by the total number of pixels. Applied to the edge images from Fig. 2, the errors read as follows:

$$nae|_{hybridSORN,11b} = 0.0181 \tag{6}$$

$$nae|_{fullSORN,15b} = 0.0287 \tag{7}$$

This metric can not determine whether the SORN implementation performs better or worse than the Integer reference, but it can show that the difference between both results is below 2% and 3%, respectively, which is in line with the visual evaluation. In order to further evaluate the different Sobel implementations, in the following section a larger number of test images is considered.

4.1 Algorithmic Evaluation with BSDS500

The Berkeley Segmentation Data Set 500 (BSDS500) [1] is a set of images for the performance evaluation of contour detection and image segmentation algorithms, consisting of images of humans, animals, objects and landscapes. For a comprehensive evaluation, the 200 test images from the data set are processed with the different Sobel implementations for all presented SORN datatypes. Additionally, two different thresholds per configuration are analyzed. For the hybrid designs, the two rightmost SORN intervals with indices $w_{sorn} - 1$ and w_{sorn} are used as thresholds. The results are compared to the corresponding Integer threshold for the reference design. For the 6 b datatype for example, the interval thresholds are $T = (200, 250]$ and $T = (250, \infty]$, the corresponding Integer thresholds are $T = 200$ and $T = 250$. For the full SORN implementation thresholds near the zero-bit are utilized and the resulting edge images are negated, in order to achieve the best performance. Therefore the equivalent threshold T_e is given, which corresponds to the compared Integer threshold.

In Table 2 the results for the mean normalized absolute error between the SORN and reference edge results are given. The utilized metric represents the mean of the nae from Eq. (5) over all test images. For both the hybrid and full SORN versions the designs utilizing a linear distributed SORN datatype perform better than the log-based versions. Furthermore, the rightmost SORN

Table 2. Mean normalized absolute error between SORN and reference integer implementation for 200 test images from BSDS500 [1].

SORN Datatype		hybrid SORN				full SORN	
		6 b lin	10 b log	11 b lin		15 b log	15 b lin
mnae	$T = w_{sorn}$	0.0659	0.1200	0.0598	$T_e = w_{sorn}$	0.1396	0.0667
	$T = w_{sorn} - 1$	0.1167	0.2323	0.0852	$T_e = w_{sorn} - 1$	–	0.0673

interval thresholds lead to the best results by means of lowest difference to the Integer reference. Compared to the results for the image in Fig. 2, given in Eq. (6)–(7), the errors are slightly higher, but still below 7%. It is mentioned again, that this metric can only measure the difference between Integer and SORN implementation. For a rating of the different designs, a third, independent reference is required.

Ground Truth Reference Comparison. For this purpose, the BSDS500 contains so-called ground truth edge results. These are human made edge detections from different human subjects [1]. For evaluating the edge detections of the different SORN implementations in comparison to the Integer reference, Fig. 4 shows the mean normalized absolute error between 6 different ground truth solutions GT and the respective edge detection results E, with the image dimensions N_x and N_y and the number of test images N_i:

$$ mnae = \frac{\sum_{i=1}^{N_i} \left(\frac{\sum_{x=1}^{N_x} \sum_{y=1}^{N_y} (GT_i(x,y) \neq E_i(x,y))}{N_x N_y} \right)}{N_i} \tag{8} $$

For the hybrid and full SORN implementations, for each datatype the threshold configuration with the best results is shown, as well as the corresponding Integer configurations. Similar to the previous evaluation, those SORN implementations utilizing linear distributed datatypes perform better than the log-based versions. For this evaluation, the linear-based SORN implementations outperform even the corresponding Integer references. As mentioned above and discussed in [15], this does not necessarily indicate that the SORN-based edge detection is better than the Integer-based for any application. Nevertheless, this evaluation on BSDS500, as well as the example in Fig. 2 show that the hybrid and full

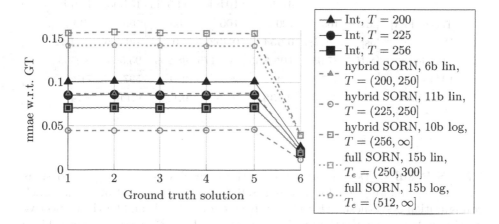

Fig. 4. Mean normalized absolute error w.r.t. 6 ground truth solutions for the different Integer and SORN Sobel implementations over 200 test images from BSDS500.

SORN-based edge detection implementations of the Sobel operator provide, at least, a similar result quality as the Integer implementation and can serve as a replacement for the SotA implementation.

4.2 Hardware Performance

In addition to the algorithmic evaluation, also the hardware performance in terms of latency, complexity and power consumption for the respective hybrid and full SORN designs, as well as for the Integer reference design is evaluated. In the following, the results of FPGA and CMOS syntheses of all designs described in Sect. 3.2 are presented.

FPGA Results. In Table 3 the synthesis results for an Artix-7 AC701 FPGA from Xilinx are given for all presented designs for a target frequency of 100 MHz. All designs are evaluated without internal pipeline registers and no DSPs are used. The worst negative slack (WNS) shows that solely the 6 b and 10 b hybrid SORN implementations are able to run at a target frequency of 100 MHz, yet all SORN designs achieve a higher maximum frequency than the Integer reference design. Concerning the required LUTs and the power consumption, the hybrid SORN approach significantly outperforms the reference design with a complexity reduction of up to 68%, whereas for the full SORN approach only the 15 b log configuration achieves a lower power consumption, all other measures can not compete with the reference design.

Table 3. FPGA synthesis results without DSPs for an Artix-7 AC701 FPGA (xc7a200tfbg676-2).

Module		Int	hybrid SORN			full SORN	
			6 b lin	10 b log	11 b lin	15 b log	15 b lin
Target Freq.	[MHz]	100	100	100	100	100	100
WNS	[ns]	−1.487	**0.554**	0.492	−0.173	−0.466	−1.042
Max Freq.	[MHz]	87.055	**105.865**	105.175	98.299	95.548	90.563
LUTs		457	**148**	207	219	597	712
Total power	[W]	0.145	**0.136**	0.137	0.138	0.140	0.147

CMOS Results. Table 4 shows the synthesis results for the proposed designs without pipeline registers for a 28 nm SOI CMOS technology from STM. Each configuration is synthesized for a target frequency of 1 GHz and for the respective maximum frequency. For the 1 GHz comparison, all SORN-based designs achieve a lower area and power consumption than the reference design, with reductions of up to 45% for area and 44% for power, respectively. Targeting maximum

Table 4. CMOS STM 28nm SOI technology synthesis results.

Module		Int	hybrid SORN			full SORN	
			6 b lin	10 b log	11 b lin	15 b log	15 b lin
Target Freq.	[MHz]	1000	1000	1000	1000	1000	1000
Runtime	[ns]	0.962	**0.958**	0.962	0.962	0.961	0.962
Area	[µm²]	1153.987	**638.765**	693.110	733.421	989.808	1132.282
Power	[µW]	550.337	329.210	349.387	349.964	**309.294**	324.075
Max. Freq.	[MHz]	1263	1681	1603	1605	1661	**1715**
Runtime	[ns]	0.792	0.595	0.624	0.623	0.602	**0.583**
Area	[µm²]	2087.165	**1100.294**	1157.251	1245.706	1661.213	2017.642
Power	[µW]	1979.710	757.566	774.962	838.914	**403.631**	413.465

frequency, all SORN-based designs achieve at minimum a 27% higher frequency than the reference design while still requiring less area and power. For this maximum frequency comparison, the hybrid SORN designs show a significantly lower area requirement than the reference design (up to 47% reduction), whereas the full SORN designs require significantly less power (up to 80% reduction).

5   Conclusion

The Sobel Operator for edge detection can be implemented as a hybrid SORN design with Integer convolution and SORN hypot function, or as a full SORN approach. Depending on the utilized SORN datatype and the chosen thresholds, both versions provide a similar algorithmic performance than the Integer reference implementation. For the presented evaluation on BSDS500 with the corresponding ground truth reference comparison, some of the SORN configurations even show a lower difference to ground truth than the Integer reference. Regarding hardware performance, the presented evaluations show that the SORN approach achieves higher frequencies and significantly lower complexity and power consumption than the Integer reference for both FPGA and CMOS.

For future work the SORN-based edge detection can be integrated into a more complex image processing system in order to provide further evaluations on the quality of the edge detection results in the context of an actual (real-time) application, for example lane detection in autonomous driving. In addition, other edge detection methods such as the Canny detector or the Marr Hildreth Operator can be implemented and evaluated for SORNs.

References

1. Arbelaez, P., Maire, M., Fowlkes, C., Malik, J.: Contour detection and hierarchical image segmentation. IEEE Trans. Pattern Anal. Mach. Intell. **33**(5), 898–916 (2011). https://doi.org/10.1109/TPAMI.2010.161

2. Bärthel, M., Rust, J., Paul, S.: Hardware implementation of basic arithmetics and elementary functions for unum computing. In: 2018 52nd Asilomar Conference on Signals, Systems, and Computers, pp. 125–129, October 2018. https://doi.org/10.1109/ACSSC.2018.8645453

3. Bärthel, M., Rust, J., Paul, S.: Application-specific analysis of different SORN datatypes for unum type-2-based arithmetic. In: 2020 IEEE International Symposium on Circuits and Systems (ISCAS), pp. 1–5 (2020). https://doi.org/10.1109/ISCAS45731.2020.9181182

4. Bärthel, M., Knobbe, S., Rust, J., Paul, S.: Hardware implementation of a latency-reduced sphere decoder With SORN preprocessing. IEEE Access 9, 91387–91401 (2021). https://doi.org/10.1109/ACCESS.2021.3091778

5. Bocco, A., Durand, Y., De Dinechin, F.: SMURF: scalar multiple-precision unum Risc-V floating-point accelerator for scientific computing. In: Proceedings of the Conference for Next Generation Arithmetic 2019, pp. 1–8 (2019)

6. Bounini, F., Gingras, D., Lapointe, V., Pollart, H.: Autonomous vehicle and real time road lanes detection and tracking. In: 2015 IEEE Vehicle Power and Propulsion Conference (VPPC), pp. 1–6 (2015). https://doi.org/10.1109/VPPC.2015.7352903

7. Cui, W., Wu, G., Hua, R., Yang, H.: The research of edge detection algorithm for Fingerprint images. In: 2008 World Automation Congress, pp. 1–5. IEEE (2008)

8. Dim, J.R., Takamura, T.: Alternative approach for satellite cloud classification: edge gradient application. Adv. Meteorol. 2013 (2013)

9. Gatopoulos, I.: Line detection: make an autonomous car see road lines. Towards Data Sci. (2019). https://towardsdatascience.com/line-detection-make-an-autonomous-car-see-road-lines-e3ed984952c

10. Glaser, F., Mach, S., Rahimi, A., Gurkaynak, F.K., Huang, Q., Benini, L.: An 826 MOPS, 210uW/MHz Unum ALU in 65 nm. In: 2018 IEEE International Symposium on Circuits and Systems (ISCAS), pp. 1–5. IEEE (2018). https://doi.org/10.1109/ISCAS.2018.8351546

11. Gustafson, J.L.: A Radical Approach to Computation with Real Numbers. Supercomput. Front. Innov. 3(2) (2016). https://doi.org/10.14529/jsfi160203

12. Gustafson, J.L.: The end of error: Unum computing. CRC Press, Boca Raton, Chapman & Hall/CRC Computational Science Series (2015)

13. Gustafson, J.L., Yonemoto, I.T.: Beating floating point at its own game: posit arithmetic. Supercomput. Front. Innov. 4(2) (2017). https://doi.org/10.14529/jsfi170206

14. Hülsmeier, N., Bärthel, M., Rust, J., Paul, S.: SORN-based cascade support vector machine. In: 2020 28th European Signal Processing Conference (EUSIPCO), pp. 1507–1511. IEEE (2021)

15. Lopez-Molina, C., De Baets, B., Bustince, H.: Quantitative error measures for edge detection. Pattern Recogn. 46(4), 1125–1139 (2013)

16. Microprocessor Standards Committee of the IEEE Computer Society: IEEE Standard for Floating-Point Arithmetic. IEEE Std. 754–2008, 1–70 (2008). https://doi.org/10.1109/IEEESTD.2008.4610935

17. Rust, J., Bärthel, M., Seidel, P., Paul, S.: A hardware generator for SORN arithmetic. IEEE Trans. Comput. Aided Des. Integr. Circuits Syst. 39(12), 4842–4853 (2020). https://doi.org/10.1109/TCAD.2020.2983709

18. Sobel, I.: An Isotropic 3×3 Image Gradient Operator. Presentation at Stanford A.I. Project 1968 (2014)

19. Solomon, C., Breckon, T.: Fundamentals of Digital Image Processing: A practical approach with examples in Matlab. Wiley, Hoboken (2011)

20. Yang, X., Yang, T.A., Wu, L.: An edge detection IP of low-cost system on chip for autonomous vehicles. In: Arabnia, H.R., Ferens, K., de la Fuente, D., Kozerenko, E.B., Olivas Varela, J.A., Tinetti, F.G. (eds.) Advances in Artificial Intelligence and Applied Cognitive Computing. TCSCI, pp. 775–786. Springer, Cham (2021). https://doi.org/10.1007/978-3-030-70296-0_56

A Posit8 Decompression Operator
for Deep Neural Network Inference

Orégane Desrentes[1,2], Diana Resmerita[2], and Benoît Dupont de Dinechin[2(✉)]

[1] ENS Lyon, Lyon, France
[2] Kalray S.A, Montbonnot-Saint-Martin, France
bddinechin@kalray.eu

Abstract. We propose a hardware operator to decompress Posit8 representations with exponent sizes 0, 1, 2, 3 to the IEEE 754 binary 16 (FP16) representation. The motivation is to leverage the tensor units of a manycore processor that already supports FP16.32 matrix multiply-accumulate operations for deep learning inference. According to our experiments, adding instructions to decompress Posit8 into FP16 numbers would enable to further reduce the footprint of deep neural network parameters with an acceptable loss of accuracy or precision. We present the design of our decompression operator and compare it to lookup-table implementations for the technology node of the targeted processor.

Keywords: Posit8 · FP16 · Deep learning inference

1 Introduction

Various approaches for reducing the footprint of neural network parameters have been proposed or deployed. Mainstream deep learning environments support rounding of the FP32 parameters to either the FP16 or BF16 representations. They support further reduction in size of the network parameters by applying linear quantization techniques that map FP32 numbers to INT8 numbers [14]. In this paper, we follow an alternate approach to FP32 parameter compression by rounding them to Posit8 numbers. As reported in [4,9], exponent sizes (es) of 0, 1, 2, 3 are useful to compress image classification and object detection network parameters ($es=2$ is now the standard for Posit8 [1]).

Unlike previous work [17,19] that apply Posit arithmetic to deep learning inference, we do not aim at computing directly with Posit representations. Rather, we leverage the capabilities of the Kalray MPPA3 processor for deep learning inference [10] whose processing elements implement 4× deep FP16.32 dot-product operators [2,3]. The second version of this processor increases 4× the number of FP16.32 dot-product operators that become 8× deep. As a result, the peak performance for FP16.32 matrix multiply-accumulate operations increases 8×. For this processor, we designed decompression operators that expand Posit8 into FP16 multiplicands before feeding them to the FP16.32 dot-product operators. These Posit8 decompression operators have to be instanced 32 times in order to match the PE load bandwidth of 32 bytes per clock cycle. The corresponding instruction provides the *es* parameter as a two-bit modifier.

In Sect. 2, we evaluate the effects of compressing the IEEE 754 binary 32 floating-point representation (FP32) deep learning parameters to Posit8 representations on classic classification and detection networks, then discuss the challenges of decompressing Posit8 representations to FP16. In Sect. 3, we describe the design of several Posit8 to FP16 decompression operators and compare their area and power after synthesis for the TSMC 16FFC technology node.

2 Compression of Floating-Point Parameters

2.1 Floating-Point Representations Considered

A floating-point representation uses a triplet (s, m, e) to encode a number x as:

$$x = (-1)^s \cdot \beta^{e-bias} \cdot 1.m, \tag{1}$$

where β is the radix, $s \in \{0, 1\}$ is the sign, $m = m_1...m_{p-1}$ is the mantissa with an implicit leading bit m_0 set to 1, p is the precision and $e \in [e_{min}, e_{max}]$ is the exponent. The IEEE-754 standard describes binary representations ($\beta = 2$) and decimal representations ($\beta = 10$). Binary representations such as FP32 and FP16 are often used in neural network inference. Let us denote the encoding of x in a representation F with x_F, so we write x_{FP32} and x_{FP16} as follows:

$$x_{\mathrm{FP32}} = (-1)^s \cdot 2^{e-127} \cdot 1.m, \text{ with } p = 24, \tag{2}$$

$$x_{\mathrm{FP16}} = (-1)^s \cdot 2^{e-15} \cdot 1.m, \text{ with } p = 11. \tag{3}$$

A first alternative to IEEE-754 floating-point for deep learning is the BF16 representation, which is a 16-bit truncated version of FP32 with rounding to nearest even only and without subnormals [13]. It has a sign bit, an exponent of 8 bits and a mantissa of 7 bits. A number represented in BF16 is written as:

$$x_{\mathrm{BF16}} = (-1)^s \cdot 2^{e-127} \cdot 1.m, \text{ with } p = 8. \tag{4}$$

A second alternative is the floating-point representation introduced by Microsoft called MSFP8 [8], which is equivalent to IEEE-754 FP16 truncated to 8 bits. This representation has a sign bit, a 5-bit exponent and a 2-bit mantissa:

$$x_{\mathrm{MSFP8}} = (-1)^s \cdot 2^{e-15} \cdot 1.m, \text{ with } p = 3. \tag{5}$$

The third alternative considered are the 8-bit Posit representations [11]. Unlike FP, Posit representations have up to four components: sign, regime, exponent and mantissa. A Posit$n.es$ representation is fully specified by n, the total number of bits and es, the maximum number of bits dedicated to the exponent. The components of a Posit representation have dynamic lengths and are determined according to the following priorities. Bits are first assigned to the sign and the regime. If some bits remain, they are assigned to the exponent and lastly, to the mantissa. The regime is a run-length encoded signed value (Table 1).

Table 1. Regime interpretation (reproduced from [5]).

Binary	0001	001	01	10	110	1110
Regime value	−3	−2	−1	0	1	2

Table 2. Comparison of components and dynamic ranges of representations. Note that the components of Posit numbers have dynamic length. The indicated values of exponent and mantissa for Posit represent the maximum number of bits the components can have. The regime has priority over the mantissa bits.

Repres	FP32	FP16	BF16	MSFP8	Posit8.0	Posit8.1	Posit8.2	Posit8.3
Exponent	8	5	8	5	0	1	2	3
Mantissa	23	10	7	2	5	4	3	2
Regime	–	–	–	–	2–7	2–7	2–7	2–7
Range	83.38	12.04	78.57	9.63	3.61	7.22	14.45	28.89

The numerical value of a Posit number is given by (6) where r is the regime value, e is the exponent and m is the mantissa:

$$x_{\text{Positn}.es} = (-1)^s \cdot \left(2^{2^{es}}\right)^r \cdot 2^e \cdot m, \text{ with } p = m. \tag{6}$$

In order to choose a suitable arithmetic representation for a set of values, one needs to consider two aspects: dynamic range and precision. The dynamic range is the decimal logarithm of the ratio between the largest representable number to the smallest one. The precision is the number of bits of the mantissa, plus the implicit one. The total size and exponent size determine the dynamic range of a given representation. Table 2 summarizes the components of the floating-point representations considered along with their dynamic range.

The FP32 representation has a wide dynamic range and provides the baseline for DNN inference. The BF16 representation preserves almost the same dynamic range as FP32, while FP16 has a smaller dynamic range and higher precision than BF16. MSFP8 has almost the same dynamic range as FP16, however it comes with a much reduced precision. Concerning Posit representations, not only they offer tapered precision by distributing bits between the regime and the following fields, but also they present the opportunity of adjusting es to adapt to the needs of a given application. An increase of the es decreases the number of bits available for the fractional part, which in turn reduces the precision.

2.2   Effects of Parameter Compression

We use pre-trained classic deep neural networks, compress their FP32 parameters to FP16 and to each of the alternative floating-point representations, after which we analyse the impact on the results of different classification and detection networks. In the following experiments, all computations are done in FP32, however we simulate the effects of lower precision by replacing the parameters with the values given by the alternative representations. Conversion from FP32 to BF16 is done by using FP32 numbers with the last 16 bits set to 0. Similarly, for the MSFP8 we use FP16 numbers with last 8 bits are cleared.

Table 3. Classification networks. Compression is applied to all parameters.

DNN	Criterion	FP32	FP16	BF16	MSFP8	Posit			
						8.0	8.1	8.2	8.3
VGG16	ACC-1	70.6	70.6	70.8	69.7	10.2	70.8	70.5	70
	ACC-5	91.3	91.3	91.2	90.3	25.2	91.0	91.0	90
VGG19	ACC-1	70.1	70.1	70.3	67.9	4.8	70.1	69.9	70.6
	ACC-5	90.4	90.4	90.5	89.4	16.3	90	90	90.4
ResNet50	ACC-1	75.7	71.3	75.5	62.8	0.0	27.7	73.2	66
	ACC-5	93.3	90.2	93.5	83.8	0.0	91.4	91.4	88.7
InceptionV3	ACC-1	71.1	71.1	71.3	44.8	65.1	69.4	69.7	63.1
	ACC-5	89.9	89.9	90.0	67.9	86.1	91.0	89.5	85.3
Xception	ACC-1	73.5	73.4	73.6	37.5	70.6	72.4	72.1	63.8
	ACC-5	92.1	92.2	91.7	60.6	90.9	91.4	90.9	86.0
MobileNetV2	ACC-1	71.2	71.2	71	0.2	12.7	12.3	11.0	3.2
	ACC-5	90.0	90.0	89.6	0.6	24.7	25.7	24.4	9.9

Table 4. Detection network. Compression is applied to all parameters.

DNN	Criterion	FP32	FP16	BF16	MSFP8	Posit			
						8.0	8.1	8.2	8.3
YOLO v3	mAP	0.41595	0.41595	0.41585	0.3022	0.4025	0.4155	0.411	0.394

Regarding the Posit representations, even at small size they encode numbers with useful precision and dynamic range. Thus, in our experiments, we evaluate Posit8 representations with *es* between 0 and 3. A dictionary containing the 255 values given by each Posit8 type is first obtained by relying on a reference software implementation [11]. We observe that all Posit8.0 and Posit8.1 values can be represented exactly in FP16. The Posit8.2 representation has 8 values of large magnitude which are not representable in FP16, but can be represented in BF16. For the Posit8.3 representation, 46 values are not representable in FP16 and 12 values are not representable in BF16. In our experiments, compression is done by replacing the parameters with their nearest values in the dictionary.

We experiment with six classification networks and one object detection network. The evaluation criteria are: Accuracy Top 1 (ACC-1), Accuracy Top 5 (ACC-5) for classification and Mean Average Precision (mAP) for detection.

Table 3 contains the results for the classification networks VGG16 [23], VGG19, ResNet50 [12], InceptionV3 [24], Xception [7], MobileNetV2 [22], which have different architectures. We also display the results obtained with FP32 and FP16 in order to compare with the standard floating-point representations. The mAP results for the object detection network (YOLO v3 [21]) are shown in Table 4.

Overall, compression with BF16 gives better results than with FP16. Despite its lower precision than FP16, BF16 appears to be well suited to deep neural network inference. On the other hand, the reduced precision of MSFP8 leads to a significant loss of performance for all tested networks. For the Posit8.0 and Posit8.3 representations, a significant loss of performance is also observed in both conventional classification (VGG16) and detection networks (YOLO).

Table 5. Classification networks. Compression is only applied to parameters of convolutions and of fully connected operators.

DNN	Criterion	FP32	Posit			
			8.0	8.1	8.2	8.3
ResNet50	ACC-1	75.7	71.3	75.0	75	73.6
	ACC-5	93.3	9.8	92.7	92.8	92.6
InceptionV3	ACC-1	71.1	66.0	70.9	70.1	69.9
	ACC-5	89.9	86.8	90.7	89.1	88.5
Xception	ACC-1	73.5	72.1	72.6	72.8	68.8
	ACC-5	92.1	91.3	91.7	91.3	89.4
MobileNetV2	ACC-1	70.8	25.3	53.5	52.7	39.4
	ACC-5	89.8	47.0	76.9	77.3	63.1

For networks containing normalization operators (ResNet50, InceptionV3, Xception and MobileNetV2), the loss of performance is significant on at least one of the Posit8 representation. This motivates a second round of experiments on the four the networks that have batch normalization operators. As reported in Table 5, not compressing the parameters of the batch normalization operators improves the performance on all these networks. However, despite the improvement, MobileNetV2 remains with a significant accuracy loss.

To summarize the effects of parameter compression to 8-bit representations, using Posit8.es with $0 \leq es \leq 3$ appears interesting. We expect that accuracy and precision could be further improved by selecting the compression representation (none, FP16, Posit8.es) individually for each operator. This motivates the design and implementation of a Posit8.es to FP16 decompression operator.

2.3   Decompression Operator Challenges

Previous implementations of Posit operators [6,15,16,18,20,25–27] include a decompressing component, often called data extraction or decoding unit, that transforms a Posit number into an internal representation similar to a floating-point number of non-standard size, into which the Posit can be exactly represented. While the structure of these units provides inspiration for our work, the design of our decompression operator faces new challenges.

- Unlike Posit operator implementations of previous work which support a single Posit representation after synthesis, our decompression operator receives the exponent size from the instruction opcode. Support of variable *es* is interesting for the MPPA3 processor as its deep learning compiler may adapt the number representation of each tensor inside a network in order to provide the best classification accuracy or detection precision.
- Posit numbers have a symmetric representation with respect to the exponent, unlike IEEE 754 floating-point that has an asymmetry tied to gradual underflow, so previous works do not deal with subnormal numbers. Support of subnormal numbers is important for the decompression of the Posit8.2 and Posit8.3 representations to FP16. The smallest Posit8.2 numbers are exactly represented as FP16 subnormals, ensuring the conversion does not underflow. Although the decompression of the Posit8.3 representation to FP16 may underflow, the range added by gradual underflow is crucial for networks that compute with FP16 parameters.
- For the Posit8 numbers with $es \in \{2,3\}$ that are not exactly representable in FP16, our decompression operator supports the four IEEE 754 standard rounding modes (Round to Nearest Even, Round Up, Round Down and Round to Zero) by extrapolating the flag setting convention described by IEEE 754 standard when narrowing to a smaller representations.
- For these Posit8.*es* number that cannot be exactly represented in FP16, our decompression operator also has to raise the IEEE 754 overflow or underflow flags. Likewise, the Not a Real (NaR) value cannot be expressed in the IEEE 754 standard, so this conversion should raise and invalid flag and return a quiet Not a Number (NaN) [1].

Lemma 1. *With the Posit8.2 and Posit8.3 representations, knowing the regime is sufficient to pre-detect conversion underflow or overflow to FP16.*

Proof. Let us call w_E the width of the floating-point exponent. For FP16 $w_E = 5$, so maximum exponent value is $e_{max} = 2^{w_E - 1} - 1 = 15$. Similarly, the smallest possible exponent (counting subnormals) is $e_{min\_sn} = -2^{(w_E - 1)} + 1 - w_M + 1 = -24$ (where w_M is the width of the mantissa).

The Posit$n$ representation combined exponent (in the floating point sense) is $c = r \times 2^{es} + e$ where r is the regime, with $-n + 1 \leq r \leq n - 2$, and e is the posit exponent $0 \leq e \leq 2^{es} - 1$. A Posit number overflows the FP16 representation when $c > e_{max}$ i.e. $r \times 2^{es} + e \geq e_{max} + 1$. This can be detected irrespective of e when $e_{max} + 1$ is a multiple of 2^{es}. This is the case when $es \leq 4$.

A Posit number underflows the FP16 representation when $c < e_{min\_sn}$. Since $c = r \times 2^{es} + e < 0$ and $e \geq 0$, then we need for the detection that $e_{min\_sn} - 1$ is a value with the maximum e, which is $2^{es} - 1$. This can be detected irrespective of e when e_{min_sn} is a multiple of 2^{es}. This is the case when $es \leq 3$.

By application of Lemma 1, Posit8.2 overflows FP16 when regime ≥ 4, while Posit8.3 overflows when regime ≥ 2 and underflows when regime ≤ -4.

Lemma 1 still holds for other values of n if the goal is to compute the IEEE 754 overflow flag, however it does not work as well for the underflow as this can only detect the underflow to zero, and not the loss of significand bits in a subnormal result. The Posit8.3 representation works here since it is small enough to not have significand bits when its value is converted to a small FP16 subnormal. The regime r should be large enough to imply no mantissa bits, i.e. $1 + r + 1 + es \geq n$, for the r corresponding to the small FP subnormals.

For example, with the FP32 representation $e_{max} = 127$ so Lemma 1 applies for conversion overflow pre-detection if $es \leq 6$. For the FP32 underflow pre-detection, $e_{min\_sn} = 154 = 2 \times 77$ so Lemma 1 does not apply in case $es > 1$.

3   Design and Implementation

3.1   Combinatorial Operator Design

Our first Posit8 decompression operator design is combinatorial (Fig 1), with steps similar to those of previously proposed Posit hardware operators. First the two's complement of the Posit number is computed when its sign bit is 1. The regime is then decoded with a leading digit counter combined with a shifter. Since the maximum exponent size es is variable, three more small shifters (Fig. 2) are used to separate the Posit exponent field e from the mantissa and to combine e with the regime value in binary to compute the unbiased FP16 exponent. The FP16 bias is then added, and the three parts of the FP16 number are combined. Additional details are needed to decompress special Posit8 numbers, or when a Posit value overflows or underflows the FP16 representation.

The special cases of Posit8 decompression to FP16 are pre-detected in the operator. Testing for Posit zero and Posit NaR are done in parallel and return floating-point zero or floating-point NaN (with the inexact flag) at the end of the operator, adding two multiplexers to the end of the operator. Those were not drawn to save space. Similarly, since all the other flag values are coded into tables, they are not drawn on the operator and not described in the figures. A few one-bit multiplexers are added to set the IEEE 754 flags.

Lemma 1 enables to check a table as soon as the regime is known to determine if the value will overflow or underflow FP16. The rest of the exponent construction is done in parallel with the table lookup. Moreover, since the mantissa of Posit8 is always smaller than the mantissa of FP16, there is no precision lost in the normal and subnormal cases. However, the rounding mode may change the values returned when there is a conversion underflow or an overflow.

The special case table (Fig. 3) outputs four different values: ∞ (0x7C00), Ω (0x7BFF), smallest subnormal (SN) (0x0001) and zero (0x0000). The sign is concatenated after the table. Two extra bits are used, the first to encode if the value in the table should be used, and the second to know if this is an overflow or an underflow, in order to use this for the flags.

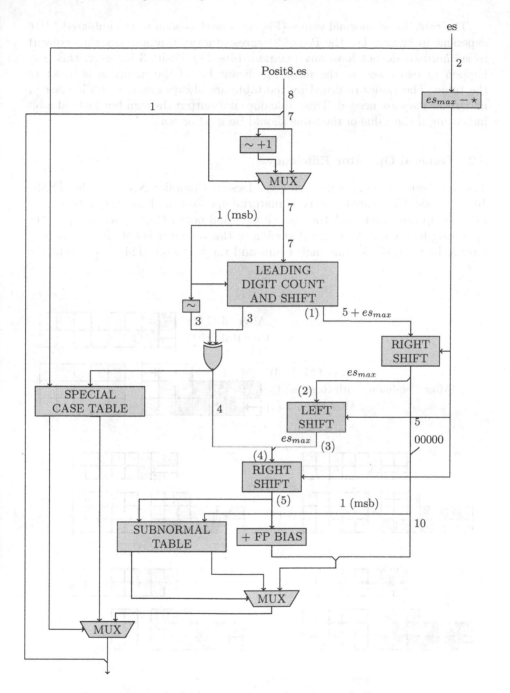

Fig. 1. Posit8 to floating-point 16 decompression operator.

The table for subnormal values (Fig. 4) is used as soon as the unbiased FP16 exponent in known. For the Posit8.2 representation, the numbers that convert to subnormals do not have any mantissa bits. For Posit8.3 however, this may happen in two cases, so the most significant bit of the mantissa is input to the table. The values returned by the table are always exact, so no inexact or underflow flags are needed. This table does not output the sign but instead a bit indicating if the value of the table should be used or not.

3.2 General Operator Efficiency

The synthesis are done with Synopsys Design Compiler NXT for the TSMC 16FFC node. We compare our combinatorial operator implementation to a baseline lookup-table and track the area (Fig. 5) and power (Fig. 6) according to the operating frequency. A pipelined version of this operator is obtained by adding a stage between the leading digit count and the first shift. This enables the use

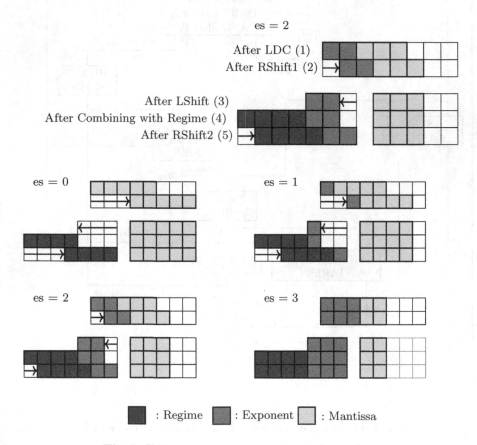

Fig. 2. Shifters to separate exponent and mantissa

sign	es	regime	rounding mode	output
0	2	4 to 6	10 or 11 (RD or RZ)	10&Ω
0	2	4 to 6	00 or 01 (RN or RU)	10&∞
1	2	4 to 6	01 or 11 (RU or RZ)	10&Ω
1	2	4 to 6	00 or 10 (RN or RD)	10&∞
0	3	2 to 6	10 or 11 (RD or RZ)	10&Ω
0	3	2 to 6	00 or 01 (RN or RU)	10&∞
1	3	2 to 6	01 or 11 (RU or RZ)	10&Ω
1	3	2 to 6	00 or 10 (RN or RD)	10&∞
0	3	-4 to -6	01 (RU)	11&SN
0	3	-4 to -6	else	11&0
1	3	-4 to -6	10 (RD)	11&SN
1	3	-4 to -6	else	11&0
*	*	*	*	00&0

exponent	mantissa msb	output
0	0	1 & 000 0010 0000 0000
0	1	1 & 000 0011 0000 0000
-1	0	1 & 000 0001 0000 0000
-1	1	1 & 000 0001 1000 0000
-2	0	1 & 000 0000 1000 0000
-3	0	1 & 000 0000 0100 0000
-4	0	1 & 000 0000 0010 0000
-5	0	1 & 000 0000 0001 0000
-6	0	1 & 000 0000 0000 1000
-7	0	1 & 000 0000 0000 0100
-8	0	1 & 000 0000 0000 0010
-9	0	1 & 000 0000 0000 0001
*	*	0 & 000 0000 0000 0000

Fig. 3. Table for the special values   **Fig. 4.** Table for the subnormal values

of a faster clock and reduces the area, the trade-off being that the conversion takes 2 clock cycles. The throughput is still of one value per clock cycle.

The lookup table implementation uses as inputs {Posit, es, rounding mode} so it has $2^{12} = 4096$ entries. The Posit sign is needed to compute the overflow and underflow IEEE 754 flags in Round Up and Round Down modes.

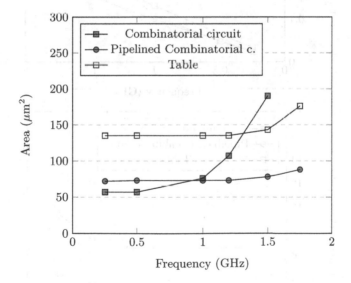

Fig. 5. Area depending on frequency for the general operators

At the target frequency of our processor (1.5 GHz), the pipelined combinatorial implementation has $\frac{2}{3}$ the area and half the leakage power of a baseline table-based implementation. It also has a lower dynamic power consumption.

3.3   Specialized Operator Efficiency

As discussed earlier, the rounding mode has no effects on the decompression for the majority of Posit8 numbers, and may only change the result in cases of overflow or underflow. This motivates specializing the operators to convert Posit8 to FP16 representations for the Round-to-Nearest (RN) mode only.

The combinatorial decompression operators are built the same way as before, the only changes being on the special case table which is significantly reduced since the rounding mode and sign no longer intervene, while the number of possible outputs is halved.

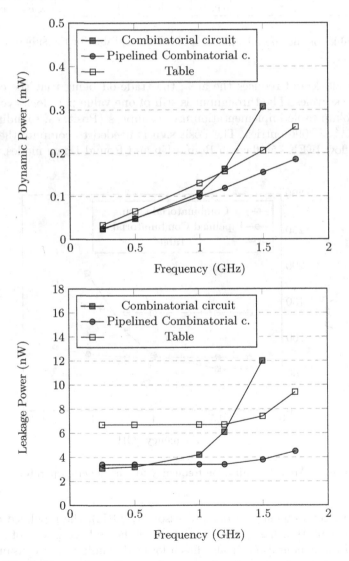

Fig. 6. Power depending on frequency for the general operators

The baseline table implementation now uses inputs {Posit, es} so it has $2^{10} = 1024$ entries. Moreover, as the sign is no longer needed for producing the IEEE 754 overflow and underflow flags, the table implementation can be factored as illustrated in Fig. 7. If the Posit is negative, its two's complement is used for accessing the table, and the sign is appended to the output of the table.

Those specialized operators show similar result in area (Fig. 8) and leakage power (Fig. 9). The factored table implementation however is significantly better

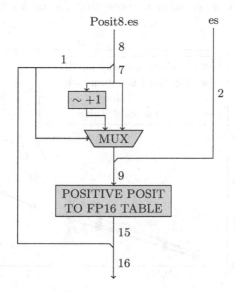

Fig. 7. Posit8 to FP16 decompression operator based on a factored table.

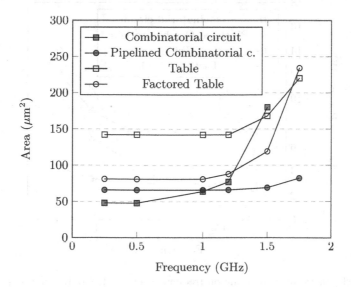

Fig. 8. Area depending on frequency for the RN only operators

than the baseline table as it exploits a symmetry that was not apparent. At our target frequency, our pipelined operator implementation has $\frac{5}{6}$ the area and $\frac{2}{3}$ the leakage power of the optimised table-based implementation, and has lower dynamic power consumption (Fig. 9).

Another specialisation of interest is to only decompress the standard Posit8.2 representation. This further reduces the size (Fig. 10) and the power consumption (Fig. 11) of the combinatorial implementations since it simplifies both the special tables and the small shifters. This operator is also smaller than the table-based implementations. It is interesting to note that for tables this small, at high

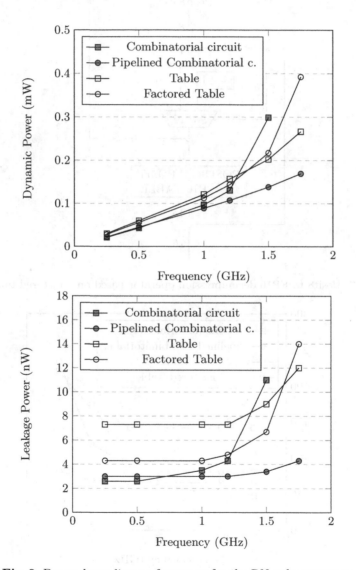

Fig. 9. Power depending on frequency for the RN only operators

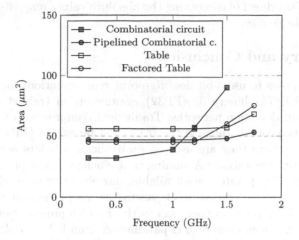

Fig. 10. Area depending on frequency for the RN and Posit8.2 only operators

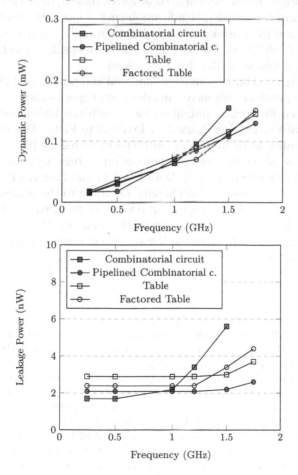

Fig. 11. Power depending on frequency for the RN and Posit8.2 only operators

frequencies the overhead of computing the absolute values may offset the benefit of halving the table size.

4   Summary and Conclusion

This paper proposes to use 8-bit floating-point representations for the compression of the IEEE 754 binary 32 (FP32) parameters of trained deep learning classification and detection networks. Traditional compression of FP32 parameters for inference rounds them to the IEEE 754 binary 16 (FP16) or to BF16 representations, where they are used as multiplicands before accumulation to FP32 or wider representations. Assuming that efficient mixed-precision FP16.32 matrix multiply-add operators are available, our objective is to select 8-bit representations suitable for floating-point parameter compression and to design the corresponding decompression operators to the FP16 representation.

We first observe that compressing parameters from FP32 to MSFP8 (a FP16 representation truncated to 8 bits proposed by Microsoft) does not give acceptable inference results for the networks considered. Indeed, to achieve compression of the FP32 parameters without significant accuracy loss, a trade-off between the dynamic range and the precision is needed. Accordingly, parameter compression to the Posit8.1, Posit8.2 and Posit8.3 representations performs well for inference with the tested networks, with a few exceptions.

We then design and implement combinatorial and table-based Posit8 to FP16 decompression operators with increasing degrees of specialization. The combinatorial designs benefit from an insight on the conditions which leads to overflow or underflow when converting Posit8.2 or Posit8.3 to FP16. This enables to pre-detect those conditions by inspecting only the Posit regime bits.

The most general decompression operators presented receive as input a Posit8 value, the exponent size $0 \leq es \leq 3$, and one of the four IEEE 754 rounding modes. A first specialization considers only rounding to the nearest even, which in turn enables the table-based implementation to be factored relative to the sign. A second specialization only decompresses the standard Posit8 representation, whose exponent size is 2. In all cases, the pipelined version of our combinatorial decompression operator appears as the best option.

References

1. Standard for Posit Arithmetic (2022) Release 5.0
2. Brunie, N.: Modified fused multiply and add for exact low precision product accumulation. In: 24th IEEE Symposium on Computer Arithmetic, ARITH 2017, London, United Kingdom, pp. 106–113, July 2017
3. Brunie, N.: Towards the basic linear algebra unit : replicating multi-dimensional FPUs to accelerate linear algebra applications. In: 2020 54th Asilomar Conference on Signals, Systems, and Computers, pp. 1283–1290 (2020)
4. Carmichael, Z., Langroudi, H.F., Khazanov, C., Lillie, J., Gustafson, J.L., Kudithipudi, D.: Performance-efficiency trade-off of low-precision numerical formats in deep neural networks. In: Proceedings of the Conference for Next Generation Arithmetic 2019, pp. 3:1–3:9. CoNGA 2019, ACM, New York (2019)

5. Carmichael, Z., Langroudi, H.F., Khazanov, C., Lillie, J., Gustafson, J.L., Kudithipudi, D.: Deep positron: a deep neural network using the posit number system. In: DATE, pp. 1421–1426. IEEE (2019)
6. Chaurasiya, R., et al.: Parameterized posit arithmetic hardware generator. In: 2018 IEEE 36th International Conference on Computer Design (ICCD), pp. 334–341. IEEE (2018)
7. Chollet, F.: Xception: Deep learning with depthwise separable convolutions. In: Proceedings of the IEEE Conference on Computer Vision and Pattern Recognition, pp. 1251–1258 (2017)
8. Chung, E., et al.: Serving dnns in real time at datacenter scale with project brainwave. IEEE Micro **38**, 8–20 (2018)
9. Cococcioni, M., Rossi, F., Ruffaldi, E., Saponara, S., de Dinechin, B.D.: Novel arithmetics in deep neural networks signal processing for autonomous driving: challenges and opportunities. IEEE Sig. Process. Mag. **38**(1), 97–110 (2020)
10. de Dinechin, B.D.: Consolidating high-integrity, high-performance, and cybersecurity functions on a manycore processor. In: 56th ACM/IEEE Design Automation Conference (DAC 2019), p. 154 (2019)
11. Gustafson, J.L., Yonemoto, I.T.: Beating floating point at its own game: posit arithmetic. Supercomput. Frontiers Innov. **4**(2), 71–86 (2017)
12. He, K., Zhang, X., Ren, S., Sun, J.: Deep residual learning for image recognition. In: Proceedings of IEEE conference on CVPR, pp. 770–778 (2016)
13. Intel: BFLOAT16 - Hardware Numerics Definition Revision 1.0, November 2018
14. Jacob, B., et al.: Quantization and training of neural networks for efficient integer-arithmetic-only inference. In: 2018 IEEE Conference on Computer Vision and Pattern Recognition, CVPR 2018, Salt Lake City, UT, USA, pp. 2704–2713, June 2018
15. Jaiswal, M.K., So, H.K.H.: Universal number posit arithmetic generator on FPGa. In: 2018 Design, Automation & Test in Europe Conference & Exhibition (DATE), pp. 1159–1162. IEEE (2018)
16. Jaiswal, M.K., So, H.K.H.: Pacogen: a hardware posit arithmetic core generator. IEEE access **7**, 74586–74601 (2019)
17. Lu, J., Fang, C., Xu, M., Lin, J., Wang, Z.: Evaluations on deep neural networks training using posit number system. IEEE Trans. Comput. **70**(2), 174–187 (2020)
18. Murillo, R., Del Barrio, A.A., Botella, G.: Customized posit adders and multipliers using the FloPoCo core generator. In: 2020 IEEE International Symposium on Circuits and Systems (ISCAS), pp. 1–5. IEEE (2020)
19. Murillo, R., Del Barrio, A.A., Botella, G.: Deep pensieve: a deep learning framework based on the posit number system. Digital Signal Process. **102**, 102762 (2020)
20. Podobas, A., Matsuoka, S.: Hardware implementation of POSITs and their application in FPGAs. In: 2018 IEEE International Parallel and Distributed Processing Symposium Workshops (IPDPSW), pp. 138–145. IEEE (2018)
21. Redmon, J., Farhadi, A.: Yolov3: an incremental improvement. CoRR abs/1804.02767 (2018)
22. Sandler, M., Howard, A.G., Zhu, M., Zhmoginov, A., Chen, L.: Mobilenetv 2: inverted residuals and linear bottlenecks. In: CVPR, pp. 4510–4520. IEEE Computer Society (2018)
23. Simonyan, K., Zisserman, A.: Very deep convolutional networks for large-scale image recognition. In: 3rd ICLR (2015)
24. Szegedy, C., Vanhoucke, V., Ioffe, S., Shlens, J., Wojna, Z.: Rethinking the inception architecture for computer vision. In: Proceedings of the IEEE Conference on Computer Vision and Pattern Recognition, pp. 2818–2826 (2016)

25. Uguen, Y., Forget, L., de Dinechin, F.: Evaluating the hardware cost of the posit number system. In: 2019 29th International Conference on Field Programmable Logic and Applications (FPL), pp. 106–113. IEEE (2019)
26. Xiao, F., Liang, F., Wu, B., Liang, J., Cheng, S., Zhang, G.: Posit arithmetic hardware implementations with the minimum cost divider and squareroot. Electronics 9(10), 1622 (2020)
27. Zhang, H., He, J., Ko, S.B.: Efficient posit multiply-accumulate unit generator for deep learning applications. In: 2019 IEEE International Symposium on Circuits and Systems (ISCAS), pp. 1–5. IEEE (2019)

Qtorch+: Next Generation Arithmetic for Pytorch Machine Learning

Nhut-Minh Ho[1]([⊠]), Himeshi De Silva[2], John L. Gustafson[1],
and Weng-Fai Wong[1]

[1] National University of Singapore, Singapore, Singapore
{minhhn,john.gustafson,wongwf}@comp.nus.edu.sg
[2] Agency for Science, Technology and Research (A*STAR), Singapore, Singapore
himeshi_de_silva@i2r.a-star.edu.sg

Abstract. This paper presents Qtorch+, a tool which enables next generation number formats on Pytorch, a widely popular high-level Deep Learning framework. With hand-crafted GPU accelerated kernels for processing novel number formats, Qtorch+ allows developers and researchers to freely experiment with their choice of cutting-edge number formats for Deep Neural Network (DNN) training and inference. Qtorch+ works seamlessly with Pytorch, one of the most versatile DNN frameworks, with little added effort. At the current stage of development, we not only support the novel posit number format, but also any other arbitrary set of points in the real number domain. Training and inference results show that a vanilla 8-bit format would suffice for training, while a format with 6 bits or less would suffice to run accurate inference for various networks ranging from image classification to natural language processing and generative adversarial networks. Furthermore, the support for arbitrary number sets can contribute towards designing more efficient number formats for inference in the near future. Qtorch+ and tutorials are available on GitHub (https://github.com/minhhn2910/QPyTorch).

Keywords: Deep Learning · Posit format · Novel number formats · Pytorch framework

1 Introduction

Reducing the bitwidth of number representations employed in Neural Networks to improve their efficiency is a powerful technique that can be used to make Deep Learning more accessible to a wider community. This is especially important when the variety of applications that use Deep Learning and the size and complexity of models have all increased drastically. For example, even with the latest GPU hardware capabilities, the GPT-3 model with 175 billion parameters requires 288 years to train [4]. The reason for the extraordinary training time and computational resources required is primarily due to the fact that the gargantuan amount of parameters cannot fit into the main memory of even the largest GPU [30]. Therefore, lowering the precision to reduce the memory consumption

J. Gustafson and V. Dimitrov (Eds.): CoNGA 2022, LNCS 13253, pp. 31–49, 2022.
https://doi.org/10.1007/978-3-031-09779-9_3

is extremely helpful to improve execution times and enable models to be run on a wider range of general-purpose hardware.

Research into low-precision number representations and their related arithmetic operations for Deep Learning has made many inroads in recent years. Several new low-precision floating-point formats have been proposed, many of them specifically targeted towards this domain. Posit^{TM} arithmetic [13] with its ability to provide tailor-made accuracy to values that are of significance in the application, has seen increasing interest. Due to the arithmetic properties of posits, they naturally lend themselves to low-precision neural network training and inference. In the case of low-precision inference, custom sets of values can also be designed for quantization to achieve high levels of model compression. Therefore, these formats merit comprehensive investigations for the use in DNN training and inference.

Due to the fast-pace and significant interest, a pressing issue the Deep Learning research community has had to grapple with in the recent past is the difficulty for independent groups to reproduce model results that are being published. Though publicly available industry benchmarks [28] have been created to address the problem, even those results cannot practically be reproduced by research groups without access to significant expertise and resources. The fine-tuning and hand-tweaked kernels are almost always proprietary and not publicly available. An open-source Deep Learning framework which enables experimenting with the aforementioned arithmetic formats, will allow researchers to quickly prototype and test newer number representations for Deep Learning.

In this paper we present Qtorch+, a framework for experimenting with posits and arbitrary number sets with flexible rounding for Deep Learning. Qtorch+ is developed upon QPyTorch, a low-precision arithmetic simulation package in PyTorch that supports fixed-point and block floating-point formats [48]. Because our framework operates seamlessly with PyTorch, users are granted all the flexibility that come with it for low-precision experimentation. This includes support for a rich set of models, benchmarks, hardware configurations and extendable APIs. Leveraging the many capabilities of Qtorch+, we evaluate an extensive set of benchmarks for both training and inference with low-precision posits and arbitrary number sets.

The remainder of the paper is organized as follows. Section 2 presents some background into Neural Networks, floating-point and fixed-point formats, posits and arbitrary number sets. It also gives an introduction into integer quantization and discusses work in the area relevant to these topics. In Sect. 3 we present the design and implementation details of the Qtorch+ framework. Section 4 gives and overview of the practical usage of the framework for training and inference. Section 5 details the results the framework achieved on inference tasks. Some case studies related to training with posits and performing inference with a customized number set are presented in Sect. 6. Section 7 concludes.

2 Background and Related Work

2.1 Neural Networks

Neural networks have achieved astonishing performance on different complex tasks in recent years. Starting with the introduction of Convolutional Neural Networks (CNN) for image classification, they have branched out to many other diverse tasks today [25]. The initial CNNs were typically trained using the back-propagation method [24] which required intensive computational power. Hardware that could handle such computational demands and the representative datasets required for training more complex tasks remained an obstacle for a long period of time. More recently, with the introduction of GPUs for accelerated training and inference at multiple magnitudes faster than traditional processors, more and more deeper neural network architectures have been designed to tackle more complex datasets and challenges (e.g. Imagenet [10]). Most notably, the introduction of very deep networks such as Resnet [14] have revolutionized the approach to computer vision with Deep Learning increasingly adopted for more difficult tasks. To this day, neural networks have been used for many tasks including vision [14,45,49], language [4,43], audio [31,37], security [36,42], healthcare [12,39], general approximation [19,47], etc.

2.2 Floating-Point and Fixed-Point Formats

Floating-point and fixed-point formats have been widely used for general computation since the early days of the computing era. They have different characteristics which make them suitable for different application domains and for different approximations. This led to various works on tuning those formats [1,6,8,11,15,18]. Recently, with the popularity of deep neural networks, hardware vendors and researchers have found that lower bitwidth on these formats can still achieve high accuracy both on inference and training while improving system energy efficiency and performance [29,40]. Thus, there are several works that target the reduced precision of floating point and fixed point format for neural network inference and training [5,40,41,44] [3,17,26,38].

Both arbitrary bitwidth floating-point and fixed-point formats have been supported by the original QPytorch framework. In this paper, we focus on extending the framework to support novel number formats such as posits and, more generally, arbitrary sets of numbers.

2.3 Integer Quantization

Integer quantization in neural networks refers to the mapping FP32 values to 8-bit integer (INT8) values. This process requires selecting the quantization range and defining the mapping function between FP32 values to the closest INT8 value and back (quantize and dequantize). If the selected range is $[\alpha, \beta]$, then uniform quantization takes an FP32 value, $x \in [\alpha, \beta]$ and maps it to an 8-bit value. The most popular mapping function used is $f(x) = s \cdot x$ (scale quantization) where

$s, x, z \in R$; s is the scale factor by which x will be multiplied. The uniform scale quantization is the most popular in hardware [46]. Let s_1 and s_2 be the scales used to quantize weight W and activation A of a dot product operation ($\otimes$). The scale quantized dot product result R' can be dequantized by multiplying with the appropriate factor:

$$R' = W' \otimes A' = \sum_1^K w_i \times s_1 \times a_i \times s_2 = R \times s_1 \times s_2$$

Integer quantization is already supported by mainstream frameworks and hardware vendors [21,46]. Thus, it is not the primary focus of this paper.

2.4 Posit Format

The posit number format has a distinctive property compared to other formats which results in better numerical stability in many application domains. The distribution of representable values in posits is more concentrated to a central point in the log2 domain (around 2^0) as seen in Fig. 1b. This property will benefit certain applications where most of the values are concentrated to a specific range. In contrast, this will overwhelm the number of representable values of both floating-point and fixed-point formats. As seen in the Figure, the floating-point accuracy distribution is uniform when compared to the tapered accuracy of posits. Consequently, many studies [7,16,22,23,27] have shown that DNNs and some specific domain applications [20] are among the beneficiaries of this property of posits.

The above described property is due to the unique representation of posits. Figure 1a shows an example of a posit. A posit environment is defined by the length of the posit, *nsize*, and the size of the exponent field, *es*, which in this case is 16 bits and 3 bits. The first bit is reserved for the sign of the number. What follows after the sign is the regime field which is of variable length. To decode the regime, one can simply count the number of 0s (or 1s) after the sign until a 1 (or 0) is reached. If the first regime bit is 0 the regime is negative and vice-versa. In this case, the regime is therefore -3. The regime value is used as the power to be raised for a value known as *useed* which is computed by using the exponent length: $(2^{2^{es}})$. There is always an implicit hidden bit in the fraction (except for zero). All these fields are read as shown in the Figure to obtain the value the posit is representing. Complete details of the posit format and its related arithmetic can be found in the posit standard [32].

2.5 Arbitrary Number Sets

Apart from the aforementioned formats, we found that allowing arbitrary number sets for inference can help accelerate the research and development of customized hardware for machine learning applications [2,34]. Thus, we also extend the framework to support any number format which can be customized depending on the application. For this feature, the user will use their own method

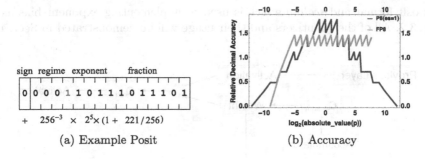

(a) Example Posit   (b) Accuracy

Fig. 1. Posit format

to craft a highly specialized table set for rounding. The arbitrary number set feature can also directly simulate any number format and other table lookup techniques. In general, any number format can be simulated with this method given the set of all representable values in the format. However, due to the table size, we recommend using this for very low bitwidth representations. As case studies of this feature, we will give some examples of using this to achieve very small sets while maintaining high output quality for selected applications. This feature will support two main research directions:

- Hardware friendly number formats with strict rules on the distribution of representable values. This category consist of number formats that are known to be efficient in multiplication (logarithmic domain, additive of logarithmic numbers).
- Arbitrary number sets which have no rules on the distribution of representable values. To implement this category in hardware, we need a customized table lookup or integer-to-integer mapping combinational logic circuit.

3   Design and Implementation of Qtorch+

Because most Deep Learning frameworks and accelerators support extremely fast FP32 inference, we can take advantage of highly optimized FP32 implementations as the intermediate form to simulate our number formats with sufficient rounding. For this to work correctly, we assume that FP32 is the superset of our target format to be simulated. This remains true when the number simulated is low bitwith (e.g. 8-bit and below). For simulating higher bitwith (above 16 bits) arbitrary number formats, we can opt to use FP64 as the intermediate number format to store the rounded values. In the context of this paper, we focus on very low bitwith number formats and using FP32 as the intermediate format. The workflow of a DNN operation simulated in a low bitwidth number format with correct rounding can be viewed in Fig. 2. This method has been widely used to simulate low precision fixed-point and floating-point formats and integer quantization in state-of-the-art techniques [48]. By introducing posits to the framework, the quantizers in Fig. 2 will have configuration parameters: *nsize, es* for posit format and *scaling* which is used to implementing exponent bias as in Sect. 3.2. All of the quantizers and their usage will be demonstrated in Sect. 4.1.

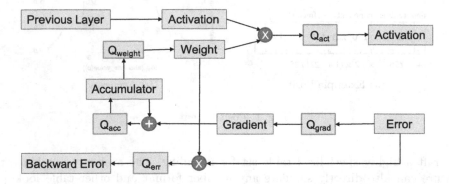

Fig. 2. Qtorch+'s training model and APIs are inherited from the original QPytorch framework with the separated kernel approach to quantize the values to new formats while using FP32 matrix multiplication for fast simulation time. We extend these functionalities to support posits, exponent biases and arbitrary number sets

3.1   Floating-Point and Posit Conversion

To simulate posits efficiently, we implement the conversion between the number format and FP32 in Qtorch+ using integer and bit operations as well as built-in hardware instructions. The implementation of the functions are based on efficient encoding and decoding of a 16 bit posit into FP32 [9].

To convert a posit into FP32, the sign bit is first extracted and the two's complement value of the number is obtained if the sign is negative. Thereafter, the regime is decoded as described in Sect. 2.4. Once these two operations are completed, we can remove these two fields with a right shift operation and directly superimpose the remaining exponent and fraction fields to the corresponding fields of an FP32 value. To get the final exponent, the decoded regime value and the exponent point bias has to be added to the exponent field.

To convert an FP32 value to a posit, first the FP32 value needs to be checked against the maximum and minimum bounds of the posit's representable range. If it can be represented as a posit, then as in the case before the sign can be extracted. The regime and exponent field of the posit can be decoded directly from the exponent field of the FP32 number. Some post-processing is done to format the regime field afterwards. Once all the fields are known, the posit can be assembled and formatted. There are many tweaks to these algorithms described that are performed to make these two operations very efficient.

3.2 Scaling and Using the Exponent Bias in Posit

After studying the the value distribution histograms of many neural networks, we found that both the weights and the activations can be scaled for a more accurate posit representation. For example, in some GANs, the weights are concentrated in the range $[2^{-4}$ to $2^{-5}]$. Therefore, we can shift the peak of the histogram to the range with highest posit accuracy, near 2^0. Note that scaling cannot provide additional accuracy for floating-point formats because their accuracy distribution is flat (see Fig. 1b).

Before and after a computation using a posit value, the encoder and decoder are used to achieve scaling. The decoder will decode the binary data in posit format to $\{S, R, E, F\}$ which represent {sign, regime, exponent, fraction}, ready for computation. The definitions of biased encoder and decoder for posit data P and a bias t are as follow:

$$\text{Biased Decoder} : \{P, t\} \rightarrow \{S, R, E - t, F\}$$
$$\text{Biased Encoder} : \{S, R, E + t, F\} \rightarrow \{P\} \tag{1}$$

We scale using the posit encoder and decoder instead of floating-point multiplications for efficiency. If we choose an integer power of 2 for the scale, input scaling and output descaling can be done by simply biasing and un-biasing the exponent value in the encoder and decoder, as shown in Eq. 1. This exponent bias can be easily implemented in hardware by additional integer adder circuit with minimal hardware cost [16].

3.3 Arbitrary Number Sets

This feature is fully supported by the extended quantizer. To use this, the user will create a full set of all possible representable values of their format and pass it as an input to the quantizer. All the real values will then be rounded to their nearest value in the given set. This feature will be described in detail and demonstrated in Sect. 6.4.

4 Practical Usage of Qtorch+

This section describes the APIs of Qtorch+ and how to use novel number formats in Deep Learning applications.

4.1 Leverage Forward_hook and Forward_pre_hook in Pytorch

To use Qtorch+ in inference seamlessly without any additional effort from the user, we leverage the "hook function" feature in Pytorch [33]. The weights can be quantized with our posit_quant function without the need for modifying the original model. However, for activation, changing the model code to intercept the dataflow of layers is required to apply custom number format simulation.

With the recent Pytorch version and the introduction of "hook" functions, there is no need to modify the original model to achieve the same result. The forward_hook function is to process the output of the current layer before going to the next layer. The forward_pre_hook function is used to process the input of the current layer before doing the layer operations. Thus, forward_pre_hook is a universal way to intercept the input of any layers while forward_hook is the convenient way to intercept the output of any layer. For general usage, we can use forward_pre_hook and preprocess activations of the current layer with low bitwidth number formats. Likewise, we use forward_hook for extra simulation of the precision of the accumulator when we do not assume the exact dot product.

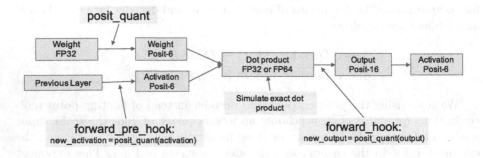

Fig. 3. Using Pytorch's feature to intercept the dataflow and simulate inference.

4.2   Qtorch+ in Training

Listing 1 shows the modification required to prepare the model for training. As we can see, the steps taken are not much different from the standard pytorch models preparation and construction. There are two main steps that we need to perform in order to use posit training:

– Declare all the quantizer used for each component of the optimizer and initialize the new optimizer with these parameters.
– Modify the model source code (MyModelConstructor) to use the argument *act_error_quant* in the forward pass of the model. The quant function must intercept the dataflow between each Convolutional/Linear layer for correct simulation. User can decide their own policy of skipping some layers to use higher precision (posit16, FP16 or FP32) if necessary.

```
1   from qtorch.quant import Quantizer, quantizer
2   from qtorch.optim import OptimLP
3   from qtorch import Posit
4   # define two different formats for ease of use
5   bit_8 = posit(nsize=8, es=2)
6   bit_16 = posit(nsize=16, es=2)
7
8   # define quantization function for each component of the neural network
9   weight_quant = quantizer(bit_8)
10  grad_quant = quantizer(bit_8)
11  momentum_quant = quantizer(bit_16)
12  acc_quant = quantizer(bit_16)
13
14  # define a lambda function so that the Quantizer module can be duplicated easily
15  act_error_quant = lambda: Quantizer(forward_number=bit_8, backward_number=bit_8)
16
17  #Step not included here: modify model forward pass to add quant() between layers.
18  model = MyModelConstrutor(act_error_quant)
19
20  #define normal optimizer as usual
21  optimizer = SGD(model.parameters(), lr=0.05, momentum=0.9, weight_decay=5e-4)
22  #user the enhanced optimizer with different number formats.
23  optimizer = OptimLP(optimizer,
24                       weight_quant=weight_quant,
25                       grad_quant=grad_quant,
26                       momentum_quant=momentum_quant,
27                       acc_quant=acc_quant,
28                       grad_scaling=2**10 ) # do loss scaling if necessary
```

Listing 1: Example of the modification needed to add to prepare the model for training with Qtorch+.

4.3   Qtorch+ in Inference

Listing 2 shows how to utilize posits (or other number formats) in inference. The code in details involve two main steps:

- Decide the number formats for processing convolutional/linear layer. It is implemented as two functions: linear_weight and linear_activation (e.g. posit(6,1) in Listing 2. Decide the number formats for processing other layers (and the layers in excluded list). This number format for other layers needs to be in high precision to prevent accuracy loss. It also needs to be compatible with the low-bitwidth format for efficient hardware design (an accelerator that supports both FP32 and posit6 is likely more expensive than the one that only support posit6 and posit16).
- Given a pretrained model, instead of looking into the model definition, we can prepare and call the prepare_model() function with the logic in Listing 2.
- In general, the simulation of the number format for output with forward_hook as in Fig. 3 can be skipped when we assume the dot product is done using the quire and the output format has enough precision to hold the output value (high precision as posit 16-bit or 32-bit).

```
1   from qtorch.quant import posit_quantize
2   def other_weight(input):
3       return posit_quantize(input, nsize=16, es=1)
4   def other_activation(input):
5       return posit_quantize(input, nsize=16, es=1)
6   def linear_weight(input):
7       return posit_quantize(input, nsize=6, es=1, scale=scale_weight)
8   def linear_activation(input):
9       return posit_quantize(input, nsize=6, es=1, scale=scale_act)
10
11  def forward_pre_hook_other(m, input):
12      return (other_activation(input[0]),)
13  def forward_pre_hook_linear(m, input):
14      return (linear_activation(input[0]),)
15
16  layer_count = 0
17  excluded_list = [] # list of all layers to be excluded from using low precision
18  model = torchvision.models.efficientnet_b7(pretrained=True) #load pretrained model
19  for name, module in model.named_modules():
20      if isinstance(module, nn.Conv2d) or isinstance(module, nn.Linear) \
21                          and layer_count not in excluded_list:
22          module.weight.data = linear_weight(module.weight.data)
23          module.register_forward_pre_hook(forward_pre_hook_linear)
24          layer_count +=1
25      else: #should use fixed-point or posit 16 bits for other layers' weight
26          if hasattr(module, 'weight'):
27              layer_count +=1
28              module.weight.data = other_weight(module.weight.data)
29              module.register_forward_pre_hook(forward_pre_hook_other)
```

Listing 2: Example of the preprocessing code needed to add to prepare the model for inference with Qtorch+. Note that this code is generic to all models which can be loaded at line 18. We do not need to modify the source code of the model definition as other frameworks. For user convenience, we can hide this whole procedure into a single function *prepare_model* which does exactly the same task.

5   Inference Results of Posit

Table 1 shows the inference results of low bitwidth posit formats on different tasks. Because our framework is fully compatible with Pytorch, we can choose a diverse set of models for difficult tasks, especially the recent state-of-the-art models [4,45,49]. Any model that has a script which can be run using Pytorch can leverage our framework. Our models include the state-of-the-art image classification model EfficientNet B7 which reaches 84.3% top 1 accuracy on Imagenet. We also include the released GPT-2 model of OpenAI which achieved state-of-the-art performance in Language Modeling when it was introduced. For performance metric, we follow the guideline of other benchmark suites which set the threshold 99% of FP32 quality when using lower precision. The vanilla posit8 (without scaling) can achieve beyond 99% accuracy of FP32 in half of the models. The accuracy of image classification models when using posit8 conforms with the 99% standard (except GoogleNet which achieves 98.9% FP32 Accuracy). The pre-trained models are retrieved from the official Pytorch[1], hugging face framework[2] and the respective authors. The inference models and scripts to run with posits are accessible online[3]. For image classification task, the test dataset is Imagenet. For Object detection, the test set is COCO 2017. For style transfer and super resolution models, we use custom datasets provided by the authors [45,49]. Question answering and language modelling task uses the SQuaD v1.1 and WikiText-103 dataset respectively.

When hardware modification is not allowed, the rest of the model can achieve 99% FP32 standard by dropping the first and the last layer of the models and apply higher precision to them (posit(16,1)). With little modification to the hardware to include an exponent bias, we can increase the accuracy of the model vastly as can be observed in column **P6+DS** in Table 1. The effect of scaling can increase the accuracy up to 7.8% in ResNEXT101. In GANs (Style Transfer and Super resolution tasks), the effect of skipping the first and the last few layers are more important than scaling posit format. Thus, we can see the P6+D can surpass posit8 in most cases. We will provide the results of posit8 when applying scaling and skipping to reach 99% FP32 standard.

6   Case Studies

6.1   Training with Posit8

Previous works have shown that posit8 is enough for training neural network to reach near FP32 accuracy both in conventional image classification application [27] and GANs [16]. In the context of this paper, we do not enhance previous results. Instead, we try to show the completeness of the framework which supports several training tasks. Because many pretrained neural networks have been fine tuned for weeks or even months, we also do not replicate the training results of these tasks. Instead, we will show a diverse training tasks on neural networks which converge in less than a day due to time constraints. For other high time-consuming tasks when training with posit, please refer to the related work which used our extension to train Generative Adversarial Networks which typically takes days to weeks to complete one experiment [16]. The results can be seen in Table 3. This training results can be reproduced with our sample code and gradient scaling configuration for **P8+** available at[4]. From the table we can see that, in contrast with inference, **P(8,2)** has dominant performance in training compared to **P(8,1)**. With correct scaling, the **P8+** can reach FP32 accuracy. This agrees with previous works [29] that gradient scaling (also known

---

[1] https://pytorch.org/vision/stable/models.html.
[2] https://huggingface.co.
[3] https://github.com/minhhn2910/conga2022.
[4] Same link as footnote 5.

Table 1. Inference results of 12 models on different tasks. For Image classification applications, the values are accuracy %. For Object Detection application, the values are box average precision (boxAP %). For GAN (style transfer and super resolution), the values are structural similarity index measure (SSIM %). For Question Answering, the values are F1 scores. For Language Modelling task, the value are Perplexity(lower better). P(6,1) means posit format with 6-bit nsize and 1-bit es. P6+D means applying the best posit 6-bit configuration while dropping (excluding) ≈ 2 layers in the original models for use in higher precision. P6+DS mean applying both dropping layers and weight/activation scale (exponent bias). The cells in **bold** font are where the configurations reach 99% FP32 quality as specified by MLPerf benchmark [35].

Task	Model	**FP32**	P(8,1)	P(8,2)	P(6,1)	P(6,2)	P6+D	P6+DS
Image classification	Resnet50	76.1	**75.7**	75.3	66.3	54.9	69.1	74.4
Image classification	ResNEXT101	79.3	**78.8**	78.4	66.3	65.8	69.8	77.6
Image classification	GoogleNet	69.8	69.0	68.8	55.8	34.9	59.4	65.5
Image classification	EfficientNetB7	84.3	**84.0**	**83.7**	79.8	75.3	80.2	82.7
Object detection	FasterRCNN	36.9	36.4	36.2	25.5	24.0	28.2	35.5
Object detection	MaskRCNN	37.9	**37.5**	37.2	36.9	25.3	28.8	36.5
Object detection	SSD	25.1	21.3	24.1	1.6	10.8	15.3	22.6
Style transfer	Horse-to-Zebra	100	96.4	93.7	84.8	79.6	98.0	**98.4**
Style transfer	VanGogh Style	100	95.0	90.6	80.7	72.3	96.2	**96.7**
Super resolution	ESRGAN	100	95.1	89.7	72.9	61.1	**99.2**	99.6
Question answering	BERT-large	93.2	**93.2**	**93.2**	92.8	92.9	92.9	92.9
Language modeling	GPT2-large ↓	19.1	**19.1**	**19.2**	21.4	22.0	20.8	19.5

Table 2. Enhancing the benchmarks in Table 1 to reach the 99% FP32 standard with posit 8-bit and with layer skipping and scaling. We pick the best accuracy among P(8,1) and P(8,2) for each model to present the result

Models	GoogleNet	FasterRCNN	SSD	Horse-to-Zebra	VanGoghStyle	ESRGAN
P8+	69.5	36.6	24.8	99.8	99.7	99.9

as loss scaling) in low precision is advantageous and should be applied as a standard procedure in other training framework. For network like VGG11, we saw that correct gradient scaling can recover the training accuracy from 22.9% to 88.6%. In this experiment, we manually set the gradient scales based on experimenting all the power-of-2 scales possible (from 2^{-10} to 2^{10}) and choose the best scale which results in the best training output. The scale is static and be used for the entire training without changes. We set the number of training epochs to be 10 epochs for Lenet and Transformer[5] and 20 epochs for other networks.

---

[5] For Transformer, we had to use P(16,2) for the backward error propagating instead of P(8,2) to achieve convergence.

Table 3. Training and inference with Qtorch+ and Posit. P8+ means the posit 8-bit configuration is used with gradient scaling that achieves highest output quality.

Model	Task (metric)	Dataset	FP32	P(8,1)	P(8,2)	P8+
Lenet	Classification (Top1 %)	MNIST	98.7	9.8	98.6	99.0
Resnet	Classification (Top1 %)	Cifar10	91.0	11.1	89.7	91.6
VGG	Classification (Top1 %)	Cifar10	87.5	10.0	22.9	88.6
Resnet	Classification (Top1 %)	Cifar100	72.9	61.8	72.4	72.7
Transformer	Translation (BLEU %)	30 k	35.4	32.9	34.5	35.0

6.2 Tips for Training with 8-Bit Posit

After trial and error, we have summarized a few tips on how to successfully train neural networks with posit8, especially with model that is difficult to train in low precision and fail to converge:

- Reduce batch size and use the built-in gradient/loss scaling. The effect of batch size and gradient scaling will be studied in this section.
- If gradient scaling still does not help convergence, the bitwidths need to be increased. Heuristically, we found that increasing the backward error precision is enough for convergence. Because the forward pass of most models is working well with posit8, they generally do not need higher bitwidth in training.
- Adjusting gradient scaling in training is generally more important than adjusting the weight scaling and exponent bias of posits.

It is a rule of thumb that using mini-batch will improve training accuracy. However, small batches mean low utilization of GPUs and longer training time. Each model has their own default batch size which is used in our experiment in Table 3. The experiment with different batch sizes are presented in Fig. 4. From the figure, we can see that the batch size parameter affects the accuracy of both FP32 and posit training. However, large batch size has stronger adverse effect on the vanilla posit format. We also conclude that, where the vanilla posit format cannot help convergence, gradient scaling must be used. In rare cases, when the gradient scaling on low bitwidth posit format still cannot help convergence, increasing the bitwidth should be considered. In the experiment in Fig. 4, we try to further use weight scaling and gradient bias similar to inference but the effect is not significant and weight-scaling/exponent bias alone cannot help posit8 training reach FP32 accuracy as gradient scaling does.

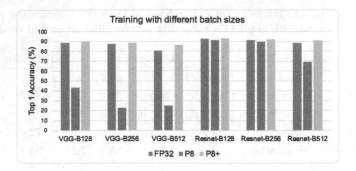

Fig. 4. Training with different batch sizes and the effect of gradient scaling. The variants used are VGG11 and Resnet18. VGG-B128 means VGG11 with 128 images in a batch.

6.3   Inference with Lower Posit Bitwidth

Section 5 shows that posit6 is still good for some inference tasks. In this section we will pick a few tasks which have high posit6 quality and further reduce the precision down to 3 bits to observe the output quality. For each bitwidth, we only select the format with the best accuracy and perform scaling and layer skipping similar to Sect. 5. The results can be seen in Fig. 5.

6.4   Inference with Arbitrary Number Set

To demonstrate the ability of the framework to support designing custom number formats, we conduct experiments with a logarithmic number format. The format is a series of power-of-two values, with the exponent represented with the fixed-point format with N bit. Let $I.F$ be a signed fixed point format with I bits signed integer and F bit fraction part. We can construct a logarithmic format based on the equation: $\pm 2^{I.F}$. The total number of bits required to represent the format is: I+F+1(sign bit). For this type of format, the multiplication is simple because it

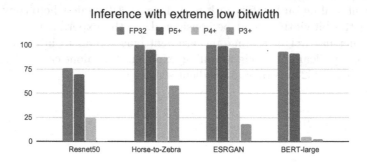

Fig. 5. Inference with bitwidth lower than 6.

can be performed by adding the I.F fixed point numbers. To use this feature, the user will generate all possible representable values of the format and supply to our quantizer (*configurable_table_quantize*). Our new quantizer will take all the representable values as an array and round real values to their nearest entry in the given numbers set. Figure 6 shows the inference results of multiple networks on the aforementioned formats. Figure 6, the $L2(7 - bit)$ means the $2^{I.F}$ format with I+F+1 = 7 bit. As we can see, a customized format can perform reasonable well on different neural networks with enough bitwidth. However, it cannot reach posit accuracy when using extreme low bitwidth (3–4 bits).

User can easily create their own format, or even a random number set without generating rules and optimized the values in the set to improve accuracy. For optimize number sets with only 4–8 distinct values but achieve good output quality on other networks (ESRGAN, GPT-2), we will have an online demonstration on our GitHub repository. Describing and implementing optimizing method for arbitrary numbers set is beyond the scope of this paper. In this section we only present the features and demonstrations.

6.5   Overhead of the Framework

Simulating number formats without hardware support will incur certain overhead on converting the format from and to the primitive FP32 format in the hardware. Our conversions are implemented both in CPU and GPU to support the variety of systems. In the end-to-end pipeline, especially in training, the overhead of simulating novel number formats is overshadowed by other time consuming tasks (data fetching, optimizer, weight update). The overhead of inference and training varies vastly between models. Measuring the computation time to complete one epoch, we got 29% slowdown when training Resnet and 81% slowdown when training VGG. However, when considering the whole end-to-end training of 20 epochs, Resnet and VGG got 21% and 70% slowdown respectively. The measured time for inference the whole Imagenet test dataset showed insignificant overhead (<10% in the models we tested). For generic models, our overhead is in range with the original QPytorch framework [48] ≈30%.

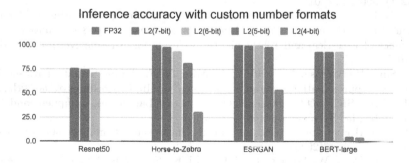

Fig. 6. Inference with a custom $2^{I.F}$ number format

7   Conclusion

We have presented the design, implementation and usage of Qtorch+, an extension to Pytorch framework to enable effortless novel number formats inference and training of neural networks. The extension is fully compatible with recent Pytorch version and therefore can be applied to many state-of-the-art models. As shown in our experiment, 8-bit posit arithmetic with scaling and kipping layers are sufficient to pass the 99% FP32 quality standard set by the community. With further extension to remove the restriction on representable number distribution, we support and arbitrary number set for use in Pytorch. This can lead to further development and research on novel low-bitwidth number formats and hardware accelerators in the near future. The tool is available open source and can also be install with *pip* package manager. At the time of writing this paper, our first version of Qtorch+ has received more than 2,000 Python package installations from around the world.

References

1. Abdelfattah, A., et al.: A survey of numerical linear algebra methods utilizing mixed-precision arithmetic. Int. J. High Perform. Comput. Appl. **35**(4), 344–369 (2021)
2. Bagherinezhad, H., Rastegari, M., Farhadi, A.: LCNN: lookup-based convolutional neural network. In: Proceedings of the IEEE Conference on Computer Vision and Pattern Recognition, pp. 7120–7129 (2017)
3. Boo, Y., Sung, W.: Fixed-point optimization of transformer neural network. In: ICASSP 2020 IEEE International Conference on Acoustics, Speech and Signal Processing (ICASSP), pp. 1753–1757. IEEE (2020)
4. Brown, T.B., et al.: Language models are few-shot learners. arXiv preprint arXiv:2005.14165 (2020)
5. Carmichael, Z., Langroudi, H.F., Khazanov, C., Lillie, J., Gustafson, J.L., Kudithipudi, D.: Performance-efficiency trade-off of low-precision numerical formats in deep neural networks. In: Proceedings of the Conference for Next Generation Arithmetic 2019, pp. 1–9 (2019)
6. Chiang, W.F., Baranowski, M., Briggs, I., Solovyev, A., Gopalakrishnan, G., Rakamarić, Z.: Rigorous floating-point mixed-precision tuning. ACM SIGPLAN Not. **52**(1), 300–315 (2017)
7. Cococcioni, M., Ruffaldi, E., Saponara, S.: Exploiting posit arithmetic for deep neural networks in autonomous driving applications. In: 2018 International Conference of Electrical and Electronic Technologies for Automotive, pp. 1–6. IEEE (2018)
8. De Silva, H., Santosa, A.E., Ho, N.M., Wong, W.F.: ApproxSymate: path sensitive program approximation using symbolic execution. In: Proceedings of the 20th ACM SIGPLAN/SIGBED International Conference on Languages, Compilers, and Tools for Embedded Systems, pp. 148–162 (2019)
9. De Silva, H.P.: Software techniques for the measurement, management and reduction of numerica (2020)
10. Deng, J., Dong, W., Socher, R., Li, L.J., Li, K., Fei-Fei, L.: ImageNet: a large-scale hierarchical image database. In: 2009 IEEE Conference on Computer Vision and Pattern Recognition, pp. 248–255. IEEE (2009)

11. Gaffar, A.A., Mencer, O., Luk, W.: Unifying bit-width optimisation for fixed-point and floating-point designs. In: 12th Annual IEEE Symposium on Field-Programmable Custom Computing Machines, pp. 79–88. IEEE (2004)
12. Gawehn, E., Hiss, J.A., Schneider, G.: Deep learning in drug discovery. Mol. Inf. **35**(1), 3–14 (2016)
13. Gustafson, J.L., Yonemoto, I.T.: Beating floating point at its own game: posit arithmetic. Supercomput. Front. Innov. **4**(2), 71–86 (2017)
14. He, K., Zhang, X., Ren, S., Sun, J.: Deep residual learning for image recognition. In: Proceedings of the IEEE Conference on Computer Vision and Pattern Recognition, pp. 770–778 (2016)
15. Ho, N.M., Manogaran, E., Wong, W.F., Anoosheh, A.: Efficient floating point precision tuning for approximate computing. In: 2017 22nd Asia and South Pacific Design Automation Conference (ASP-DAC), pp. 63–68. IEEE (2017)
16. Ho, N.M., Nguyen, D.T., Silva, H.D., Gustafson, J.L., Wong, W.F., Chang, I.J.: Posit arithmetic for the training and deployment of generative adversarial networks. In: 2021 Design, Automation Test in Europe Conference Exhibition (DATE), pp. 1350–1355 (2021). https://doi.org/10.23919/DATE51398.2021.9473933
17. Ho, N.M., Vaddi, R., Wong, W.F.: Multi-objective precision optimization of deep neural networks for edge devices. In: 2019 Design, Automation and Test in Europe Conference and Exhibition (DATE), pp. 1100–1105. IEEE (2019)
18. Ho, N.M., Wong, W.F.: Exploiting half precision arithmetic in Nvidia GPUs. In: 2017 IEEE High Performance Extreme Computing Conference (HPEC), pp. 1–7. IEEE (2017)
19. Ho, N.M., Wong, W.F.: Tensorox: accelerating GPU applications via neural approximation on unused tensor cores. IEEE Trans. Parallel Distrib. Syst. **33**(2), 429–443 (2021)
20. Klöwer, M., Düben, P.D., Palmer, T.N.: Posits as an alternative to floats for weather and climate models. In: Proceedings of the Conference for Next Generation Arithmetic 2019, pp. 1–8 (2019)
21. Krishnamoorthi, R., James, R., Min, N., Chris, G., Seth, W.: Introduction to quantization on PyTorch (2020). https://pytorch.org/blog/introduction-to-quantization-on-pytorch/
22. Langroudi, H.F., Carmichael, Z., Gustafson, J.L., Kudithipudi, D.: Positnn framework: tapered precision deep learning inference for the edge. In: 2019 IEEE Space Computing Conference (SCC), pp. 53–59. IEEE (2019)
23. Langroudi, H.F., Karia, V., Gustafson, J.L., Kudithipudi, D.: Adaptive posit: parameter aware numerical format for deep learning inference on the edge. In: Proceedings of the IEEE/CVF Conference on Computer Vision and Pattern Recognition Workshops, pp. 726–727 (2020)
24. LeCun, Y., et al.: Backpropagation applied to handwritten zip code recognition. Neural Comput. **1**(4), 541–551 (1989)
25. LeCun, Y., et al.: Lenet-5, convolutional neural networks. **20**(5), 14 (2015). http://yann.lecun.com/exdb/lenet
26. Lo, C.Y., Lau, F.C., Sham, C.W.: Fixed-point implementation of convolutional neural networks for image classification. In: 2018 International Conference on Advanced Technologies for Communications (ATC), pp. 105–109. IEEE (2018)
27. Lu, J., Fang, C., Xu, M., Lin, J., Wang, Z.: Evaluations on deep neural networks training using posit number system. IEEE Trans. Comput. **70**(2), 174–187 (2020)
28. Mattson, P., et al.: MLPerf: an industry standard benchmark suite for machine learning performance. IEEE Micro **40**(2), 8–16 (2020)

29. Micikevicius, P., et al.: Mixed precision training. arXiv preprint arXiv:1710.03740 (2017)
30. Nvidia: Scaling Language Model Training to a Trillion Parameters Using Megatron. https://developer.nvidia.com/blog/scaling-language-model-training-to-a-trillion-parameters-using-megatron/ (2021). Accessed 03 Jan 2022
31. Oord, A.V.D., et al.: WaveNet: a generative model for raw audio. arXiv preprint arXiv:1609.03499 (2016)
32. Posithub.org: Posit Standard Documentation Release 3.2-draft. https://posithub.org/docs/posit_standard.pdf (2018). Accessed 03 Jan 2022
33. Pytorch: Pytorch module (2021). https://pytorch.org/docs/stable/generated/torch.nn.Module.html#torch.nn.Module.register_forward_pre_hook
34. Ramanathan, A.K., et al.: Look-up table based energy efficient processing in cache support for neural network acceleration. In: 2020 53rd Annual IEEE/ACM International Symposium on Microarchitecture (MICRO), pp. 88–101. IEEE (2020)
35. Reddi, V.J., et al.: MLPerf inference benchmark. In: 2020 ACM/IEEE 47th Annual International Symposium on Computer Architecture (ISCA), pp. 446–459. IEEE (2020)
36. Ryan, J., Lin, M.J., Miikkulainen, R.: Intrusion detection with neural networks. Adv. Neural Inf. Process. Syst. 943–949 (1998)
37. Sigtia, S., Benetos, E., Dixon, S.: An end-to-end neural network for polyphonic piano music transcription. IEEE/ACM Trans. Audio Speech Lang. Process. 24(5), 927–939 (2016)
38. Solovyev, R., Kustov, A., Telpukhov, D., Rukhlov, V., Kalinin, A.: Fixed-point convolutional neural network for real-time video processing in FPGA. In: 2019 IEEE Conference of Russian Young Researchers in Electrical and Electronic Engineering (EIConRus), pp. 1605–1611. IEEE (2019)
39. Sordo, M.: Introduction to neural networks in healthcare. Knowledge Management for Medical Care, Open Clinical (2002)
40. Sun, X., et al.: Hybrid 8-bit floating point (HFP8) training and inference for deep neural networks (2019)
41. Sun, X., et al.: Ultra-low precision 4-bit training of deep neural networks. Adv. Neural. Inf. Process. Syst. 33, 1796–1807 (2020)
42. Tobiyama, S., Yamaguchi, Y., Shimada, H., Ikuse, T., Yagi, T.: Malware detection with deep neural network using process behavior. In: 2016 IEEE 40th Annual Computer Software and Applications Conference (COMPSAC), vol. 2, pp. 577–582. IEEE (2016)
43. Vaswani, A., et al.: Attention is all you need. In: Advances in Neural Information Processing Systems, pp. 5998–6008 (2017)
44. Wang, N., Choi, J., Brand, D., Chen, C.Y., Gopalakrishnan, K.: Training deep neural networks with 8-bit floating point numbers. In: Proceedings of the 32nd International Conference on Neural Information Processing Systems, pp. 7686–7695 (2018)
45. Wang, X., et al.: ESRGAN: enhanced super-resolution generative adversarial networks. In: Proceedings of the European Conference on Computer Vision (ECCV) Workshops (2018)
46. Wu, H., Judd, P., Zhang, X., Isaev, M., Micikevicius, P.: Integer quantization for deep learning inference: principles and empirical evaluation. arXiv preprint arXiv:2004.09602 (2020)
47. Yazdanbakhsh, A., Park, J., Sharma, H., Lotfi-Kamran, P., Esmaeilzadeh, H.: Neural acceleration for GPU throughput processors. In: Proceedings of the 48th International Symposium on Microarchitecture, pp. 482–493 (2015)

48. Zhang, T., Lin, Z., Yang, G., De Sa, C.: QPyTorch: a low-precision arithmetic simulation framework. arXiv preprint arXiv:1910.04540 (2019)
49. Zhu, J.Y., Park, T., Isola, P., Efros, A.A.: Unpaired image-to-image translation using cycle-consistent adversarial networks. In: Proceedings of the IEEE International Conference on Computer Vision, pp. 2223–2232 (2017)

ACTION: Automated Hardware-Software Codesign Framework for Low-precision Numerical Format SelecTION in TinyML

Hamed F. Langroudi[✉], Vedant Karia, Tej Pandit, Becky Mashaido, and Dhireesha Kudithipudi

Neuromorphic AI Lab, University of Texas at San Antonio, San Antonio, TX, USA
seyedhamed.fatemilangroudi@utsa.edu

Abstract. In this paper, a new low-precision hardware-software codesign framework is presented, to optimally select the numerical formats and bit-precision for TinyML models and benchmarks. The selection is performed by integer linear programming using constraints mandated by tiny edge devices. Practitioners can use the proposed framework to reduce design costs in the early stages of designing accelerators for TinyML models. The efficacy of various numerical formats is studied within a new low-precision framework, ACTION. Results assert that generalized posit and tapered fixed are suitable numerical formats for TinyML when the trade-off between accuracy and hardware complexity is desired.

Keywords: Deep neural networks · Low-precision arithmetic · Hardware-Software Codesign

1 Introduction

TinyML is an emerging machine learning (ML) field that aims to bring intelligence on ubiquitous tiny edge platforms with ≤ 1 MB memory footprint, 100 MOPS (million operations per second) throughput, and ≤ 1 mW power consumption [1]. The capability to perform ML inference on edge devices enabled by TinyML, can expand the scope of ML applications to new areas such as nature conservation [2], and STEM education [3]. Moreover, the on-device inference capabilities provided by TinyML bypass the latency and energy consumption of data transition between the device and cloud to enhance privacy and security. However, the resource limitations of edge devices introduce significant challenges to perform on-device ML inference on current TinyML models with thousands of parameters and millions of computations [4].

Quite often, to deploy TinyML models on tiny edge devices, the ML inference is performed with low-precision numerical formats [5–11]. The low-precision numerical format offers complexity reduction in multiple dimensions, such as computational resources, energy and memory footprint [5,11]. However, the benefits of

H.F. Langroudi and V. Karia—Equal contribution.

low-precision numerical format come at the expense of model performance [12]. The trade-off between hardware complexity and accuracy loss differs between different numerical format configurations [13]. The non-uniform numerical format such as posit [14] has better accuracy and more hardware complexity compared to a hardware-oriented and equispaced numerical format such as fixed-point. This incongruence in accuracy and hardware complexity offered by various numerical formats introduces a broad and large design space for numerical format exploration. Tangential to this, the hardware and model performance constraints are varied from one edge device to another. The process of manual selection of numerical format is ad-hoc and sub-optimal due to the large design exploration space and variability in constraints. Therefore, the process of selecting the optimal numerical format for a TinyML target requires an automatic hardware-software co-design framework that considers model performance and hardware complexity constraints. Such a framework can be used by practitioners and startups as an Early-DSE [15] (early stage design space exploration) framework that generates the template for a suitable accelerator, including the numerical format specification (to aid in reducing the cost of the accelerator's design). Other frameworks that automatically select an appropriate low-precision numerical format based on constraints that are mandated by TinyML model performance and tiny edge platform limitations have been previously proposed in literature [5–7,11]. However, the scope of these existing hardware-software co-design frameworks have been limited to the selection of a bit-precision of the fixed-point numerical format for a particular layer of a TinyML model [5,6]. Moreover, the current frameworks to select a low-precision numerical format use computationally intensive reinforcement learning (RL) algorithms with high sensitivity to initial parameter selection [5,6].

Therefore, we propose a hardware-software co-design framework, called ACTION, that finds the optimal numerical format configuration through integer linear programming (ILP) inspired from recent studies in mixed-precision quantization [16,17]. Using ILP optimization instead of RL reduces the search time, bypasses the need for hyperparameter optimization, and reduces computational overheads [16]. Specifically, the optimal numerical format configuration achieved through the ILP solver minimizes or maximizes one of the objective metrics (e.g., accuracy) while the other subjective constraints (e.g., latency, memory footprint) are met. Unlike the existing frameworks, ACTION, supports a broad range of numerical formats including posit and generalized posit, summing up to a total of 60 possible numerical format configurations.

The key contributions of this work are as follows:

1. We develop a low-precision hardware-software co-design framework to constrain the early stage design space exploration which selects an appropriate numerical format based on the custom user defined constraints through integer linear programming optimization.
2. Various configurations and dataflows in a systolic array based architecture are studied to evaluate the performance of the numerical formats when incorporated in an accelerator.

2 Background

A non-zero finite real number y is represented by Eq. (1) where s is the sign, L is the bit array, ϕ is a function mapping the bit array $f \in [0, 1)$ as a fraction to a real value, and $\star$ is an arbitrary function between the integer and the fraction (in this study $\star \in \{\times, +\}$)

$$y = (-1)^s \times \psi(L) \star \phi(f) \tag{1}$$

The numerical format used in this study is summarized in Table 1, based on Eq. (1). Note that all numerical formats use a two's complement representation to represent a negative number except for the floating point numerical format, which uses a sign-magnitude representation. The main difference between these numerical formats is the way that the bit array L is encoded. In traditional numerical formats such as fixed and floating point, L is binary(B) and offset-binary(OB) encoded respectively while in recent numerical formats such as tapered fixed-point (taper [18]), L is signed unary encoded or regime encoded (RE) where the *runlength* m of identical bits $(l...l)$ is terminated by either an compliment bit $\bar{l}$ where $m \leq n$ or by a final bit. Hence the value R in regime encoding is computed as (2).

$$R = \begin{cases} -m, & l = 0 \\ m - 1, & l = 1 \end{cases} \tag{2}$$

In posit and generalized posit numerical formats the bit string l is divided into two parts, the regime and the exponent. The regime bit array is singed unary encoded and the exponent bit array is binary encoded. The signed unary encoding is a variable encoding that adds a tapered accuracy attribute to the posit, generalized posit and tapered fixed-point formats. In numerical formats with **tapered-accuracy**, the density of values is highest near 0 and then tapers towards the maximum-representable number as shown in Fig. 1.

Table 1. Description of numerical formats that are explored in this study.

Format	L Encode	$\psi(L)$	$\star$	$\phi(f)$	Parameters
Fixed-point	B	L	$+$	f	–
Tapered fixed-point	RE	R	$+$	f	Is, sc
Floating point	OB	$2^{e-2^{es-1}-1}$	$\times$	$1+f$	–
Posit	RE, B	$2^{2^{es}R+e}$	$\times$	$1+f$	–
Generalized posit	RE, B	$2^{2^{es}R+e+e_b}$	$\times$	$1+f$	rs, e_b

The numerical formats with a tapered-accuracy characteristic are more appropriate to represent TinyML model parameters (weights) due to their bell-shaped distribution [9].

Among numerical formats with the tapered-accuracy characteristic, only generalized posit and tapered fixed-point can accommodate the variability observed in a layer's parameter distribution by assigning two additional hyperparameters that can modify it's dynamic range and tapered precision [7, 19]. Ordinarily, the maximum accuracy is located at 1 in posit, generalized posit and tapered fixed-point formats. The Exponent bias (e_b) and scaling factor (sc) can re-center the location of maximum accuracy from 1 to 2^{e_b} or 2^{sc}. The dynamic range and shape of the numerical format values' distribution (maximum tapered to uniform) is controlled by a maximum regime/integer run-length (rs/Is) parameter.

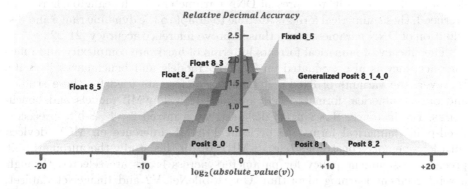

Fig. 1. The relative decimal accuracy [14] for various 8-bit numerical formats Float 8_5 , Float 8_4, Float 8_3 are 8-bit floating format with 5, 4 and 3 exponent bits, respectively, and Posit 8_0, Posit 8_1, and Posit 8_2 are 8-bit posit format with 0, 1, and 2 exponent bits respectively. The Fixed 8_5 indicates fixed-point numerical format with 5-bit integer and 3 fraction bits, and Generalized posit 8_1_4_0 is 8-bit generalize posit numerical format with $es = 1$, $rs = 4$, and $e_b = 0$.

3 Related Work

In recent years, the impact of various low-precision numerical formats on deep learning inference accuracy has been studied thoroughly [12, 13, 20–22]. For instance, Gysel et al. proposed the Ristretto framework to explore the effect of fixed-point, minifloat (8-bit floating point format with arbitrary exponent and fraction bit-width), and block floating point (where each block of floating point numbers used a shared exponent) on classification accuracy [20]. The outcome of this study on the CIFAR-10 corpus shows negligible accuracy difference between DNN inference with an AlexNet model using 8-bit and 32-bit floating point format parameters.

However, a few works proposed empirical frameworks that demonstrate the effect of numerical formats on the trade-off between performance and hardware complexity. For instance, Hashemi et al. demonstrate that DNN inference with 8-bit fixed-point using AlexNet (on CIFAR-10 dataset) results in a 6.8× improvement in energy consumption with <2% accuracy degradation compared to a

DNN inference with 32-bit floating point [13]. Following this work, Langroudi *et al.* introduce the Cheetah framework where the trade-off between inference accuracy and hardware complexity (e.g., energy-delay product (EDP)) is provided for DNN inference with [5, 8] precision posit, float and fixed-point formats [12]. The optimal bit-precision for each numerical format in this framework is obtained through a top-down iterative process where the accuracy and hardware complexity achieved by a numerical format is compared with the specified design constraints provided by practitioners. Through this study, posit shows better accuracy and EDP trade-off as compared to float and fixed-point numerical formats. Recently, Thierry Tambe *et al.* [22] and Langroudi *et al.* [21] introduce two novel numerical formats (adaptive float and adaptive posit) to represent DNN parameters. With negligible hardware overhead, these numerical formats are able to adapt to the dynamic range and distribution of DNN parameters and thus improve inference accuracy [21, 22].

The efficacy of numerical formats in terms of hardware complexity and inference accuracy is also evaluated on TinyML models and benchmarks [5–8, 10]. However, the variants of fixed-point numerical formats used for these studies and other numerical formats is not evaluated on TinyML models and benchmarks. For instance, Rusci *et al.* demonstrate a mixed 2-, 4-, 8-bits precision fixed-point numerical format to perform TinyML inference on MCU devices with low-memory constraints (e.g., 2 MB) [6]. In this study, the automatic bit precision assignment policy for parameters across layers are selected through a reinforcement learning algorithm. On MobileNet V2 and ImageNet dataset, the aforementioned mixed-precision quanization approach results in about 1.3% inference accuracy degradation as compared to inference accuracy with 32-bit floats. Recently, Langroudi *et al.* introduce an efficient method of quantizing TinyML models using a novel tapered fixed-point numerical format that leverages the benefit of both posit (in terms of accuracy performance) and fixed-point (in terms of hardware efficiency) [7]. The tapered fixed-point has shown better EDP and accuracy trade-off over fixed-point on various benchmarks [7].

This research proposes the ACTION, framework for TinyML models where the numerical format and bit-precision of model parameters is automatically selected through ILP optimization. A notable difference between this work and previous works is that the ACTION framework supports a broad range of numerical formats and its search space exploration time is an order of magnitude faster than previous RL approach [6].

4   ACTION Framework

The goal of ACTION framework is to automatically and swiftly select the appropriate numerical format based on constraints required by TinyML benchmarks and tiny edge devices. This platform can be generalized for use on other DNN models and edge devices since it provides the ability for practitioners to choose their own constraints. This framework comprises of four key aspects as shown in Fig. 2: User Interface, Initialization, Optimizer, and Evaluator.

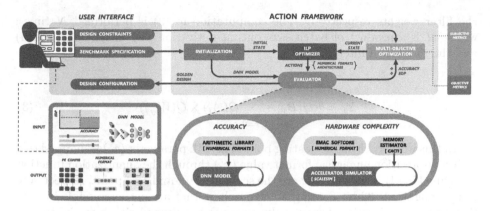

Fig. 2. The ACTION high-level low-precision hardware & software co-design framework for TinyML models on tiny edge platforms

4.1 User Interface

The goal of the user interface is to preselect the metrics and parameters given as input to the framework. These include benchmark specifications, models, datasets (e.g., TinyML v0.5 benchmark), metrics, constraints, and variables (summarized in Table 2, and 3). The framework then generates specifications of the accelerator such as numerical format configuration, PE configurations, memory requirements and data flows, that are summarized in an output file. The input and output of the user interface are specified in YAML format.

4.2 Initialization

In the initialization step, the model is trained with 32-bit floating point values. The high-precision 32-bit floating point trained weights and activations are transferred to the evaluator. The specification of the TinyML model that is used by the accuracy and hardware complexity evaluator is automatically generated.

4.3 Evaluator

The Evaluator unit is explained with the help of an example of a single hidden layer convolutional neural network to highlight its key components and operation clearly, although it can be generalized for any TinyML model.

Software Design and Exploration: A computational node in a single hidden layer convolutional neural network computes (3) where B indicates the bias vector, W is the weight tensor with numerical values that are associated with each connection, A represents the activation vector as input values to each node, θ is the activation function, Q denotes the quantization function, Y is a feature vector consisting of the output of each node, and M is equal to the product of (C, R, S) filter parameters: the number of filter channels, the filter heights, and the filter weights respectively. The computation in (3) is performed N times, where N is a product of batch size, output activation size (height and width) and the number of filters.

$$Y_j = \theta(Q(B_j) + \sum_{i=0}^{M} Q(A_i) \times Q(W_{ij})) \tag{3}$$

In this work, each 32-bit floating point TinyML parameter (x_i) is mapped to a l-bit low-precision numerical format value (x_i') through the quantization function as defined in (4), where s and z are the scaling factor and zero point, respectively. Large magnitude 32-bit floating point numbers that are not expressible in $[l, u]$ (low-precision numerical format values range) are clipped either to the format lowerbound (l) and upperbound (u). Moreover, the clipped values that lie in interval $[a, b]$ (the two consecutive low-precision numerical format values) are rounded to the nearest even number.

$$x_i' = Q(x_i, q, l, u, s, z) = \text{Round}(\text{Clip}(s \times x_i + z, l, u)) \tag{4}$$

The product of quantized activations and weights are computed with a low-precision numerical format multiplier without rounding the end products. The products are then accumulated over wide signed fixed-point register, the *quire* [14]. Note that this MAC operation for m operands is error free since quire size, as shown in (5), is selected in a way that the dynamic range of partial accumulated values are captured. The D_l in (5) represents the dynamic range of the low-precision numerical format.

$$w_{quire} = \lceil \log_2(m) \rceil + 2 \times \lceil \log_2(D_l) \rceil + 2 \tag{5}$$

Hardware System Design and Architecture: To evaluate the area, energy and latency of the hardware accelerator, analyzing only the hardware complexity introduced by various numerical formats can be misleading. Existing frameworks that evaluate the performance of the numerical formats compare only the energy consumption of the MAC operations which overlooks the constraints imposed by memory and dataflow in the accelerator.

In order to evaluate the area, and power of an accelerator which incorporates a particular numerical format, we limit the evaluation to an architecture comprising of Processing Elements (PE) arranged in a 2D systolic array configuration. The compute efficiency of the systolic array architectures depends highly on the dataflow and the PE array size since the matrix multiplication operation of a TinyML model is mapped to the PEs arranged in the 2D matrix structure. To estimate the latency of the system we bridge our framework with the SCALE-Sim tool [23] by simulating the TinyML models for various configurations of PE and dataflows as illustrated in Fig. 3.

To analyze the energy consumption and the area of the system, each PE is replaced with the multiply and accumulate (MAC) unit of the numerical formats stated in Table 3. The MAC units for various numerical formats and different configurations were synthesized on the Synopsys 32 nm CMOS technology node. Power consumption and the area of individual MAC units were combined with the cycle count and memory access details obtained from SCALE-Sim to analyze the hardware complexity of the accelerators with various incorporated numerical format configurations.

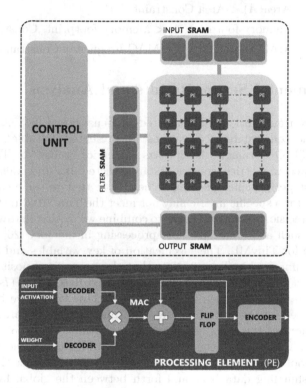

Fig. 3. The overview of the accelerator's system architecture used to evaluate the efficacy of the numerical formats on accelerators.

4.4 Optimizer

Integer Linear Programming (ILP): The ILP in this work is defined as (6) where y_i and x_i are objective and subjective metrics selected from Table 2, A as configuration sets from Table 3, C_j are constraints with respect to subjective x_j.

$$\min_{y_i} \quad \sum_{i=0}^{k=1} y_i(A) \tag{6}$$
$$\text{s.t.} \quad x_j(A) < c_j, 0 < j < 6$$

For instance the accuracy degradation (ACC_d) is selected as an objective, and area, EDP, memory footprint, and MAC frequency are chosen as subjectives, the (6) as in (7). In this study, we set the maximum number of subjective metrics to 4. On occasion, the subjective metrics have some overlap, such as power and EDP.

$$
\begin{aligned}
\min_{ACC_d} \quad & ACC_d(A) \\
\text{s.t.} \quad & \text{EDP}(A) < \text{EDP Constraint} \\
& \text{Area}(A) < \text{Area Constraint} \\
& \text{Memory footprint}(A) < \text{Memory footprint. Constraint} \\
& \text{MAC Frequency}(A) < \text{MAC Frequency Constraint}
\end{aligned}
\tag{7}
$$

5   Experimental Setup, Results and Analysis

The ACTION framework is implemented in TensorFlow [24]. A summary of the metrics and constraints specifications are presented in Table 2. The current version of ACTION framework supports 8 metrics crucial for TinyML applications. For specific metrics, constraints are selected at 3 intervals between the best and worst performance yielding values. Note that in some cases, the best possible result for a specific metric may not meet the TinyML target. This shows that the low-precision arithmetic needs to combine with other hardware/software optimizations such as pruning [25] and processing in memory [26] to meet that specific metric for TinyML. The specification of key variables and their configurations are summarized in Table 3. Note that the generalized posit hyperparameters ($rs \in [1..n - 1]$ and $e_b \in [-3, 3]$) and tapered fixed-point ($Is \in [1..n]$ and $sc \in [-3, 3]$) are not mentioned in Table 3 since these values are fixed and predetermined based on the dynamic range and distribution of parameters [7,19]. The specifications of the tasks and inference performance with 32-bit floats are summarized in Table 4. To estimate latency, we bridge our framework with the SCALE-Sim tool [23]. SCALE-Sim, however, does not consider the cycles consumed while shuttling data back and forth between the global buffer and the DRAM. Therefore, the total latency is re-approximated by considering PE array execution time and DRAM access time (Micron MT41J256M4). For the energy estimation analysis, execution time, and power consumption, we consider the use of the 32-nm CMOS technology node.

5.1   Numerical Formats' Performance on TinyML Benchmark

The Table 5 are summarized the performance of various numerical formats on TinyML v0.5 benchmark that evaluated using ACTION framework. Amongst the evaluated numerical formats, generalized posit shows the best performance. For instance, the inference accuracy on the image classification benchmark using generalized posit is improved by an average of 6.70%, 19.92%, 8.66%, 30.02%, as compared to posit, float, tapered fixed-point and fixed-point respectively. The

Table 2. The metrics and constraints specification.

Categ	Metrics	Constraints	TinyML target
1	EDP	[mean-std,mean,mean+std]	–
	Energy		–
	Power		$\leq 1\,\mathrm{mw}$ [1]
2	MAC Frequency		10–100 MHz [27]
3	PE utilization		–
4	Area		$< 20\ \mathrm{mm}^2$ [27]
6	Accuracy Degradation		1–6% [1]
7	Memory footprint		$\leq 100\,\mathrm{KB} + 0.5\,\mathrm{MB}$ [28]

Table 3. The key variable specification (P:Posit, FP:Floating point, FX: Fixed-point, GP: generalized posit, and TFX: tapered fixed-point).

Variable	Configuration	Search space
Formats	$P(n \in [5..8],\ es = [0..2])$	60
	$FP(n \in [5..8],\ e = [3..n-2])$	
	$FX(n \in [5..8],\ f = [1..n-1])$	
	$GP(n \in [5..8],\ es = [0..2])$	
	$TFX(n \subset [5..8])$	
PE	$32 \times 32,\ 32 \times 16,\ 16 \times 16,\ 12 \times 14,\ 8 \times 8$	5
Data-flow	OS, WS, IS	3
Total	–	900

Table 4. The TinyML v0.5 [4] models and benchmarks using 32-bit float parameters description.

Application	Dataset	DNN Model	# Parameters	# Ops	Performance
Keyword spotting	Speech commands v2	DS-CNN	24.91 K	5.54 M	92.15%
Visual wake words	VWW dataset	MobileNetV1	221.79 K	15.69 M	82.72%
Image classification	CIFAR10	ResNet-8	78.67 K	25.27 M	86.26%

high performance of the generalized posit numerical format on TinyML benchmarks can be credited to the capability of this numerical format to auto-adjust to the dynamic range and distribution of the weights and activations. Moreover, we observed that the performance of tapered fixed-point is not only better than fixed-point, but also, on average, comparable with floats and posit formats, which has not been previously observed [7]. This finding emphasizes that tapered fixed-point is a good candidate for TinyML models and applications. Moreover, as the number of bits is decreased to 7-bits and below, the float, fixed-point and tapered fixed-point formats show poor accuracy performance. This can be attributed to

Table 5. The TinyML inference performance using various numerical formats on TinyML v0.5 benchmark (P:Posit, FP:Floating point, FX: Fixed-point, GP: Generalized posit, and TFX: Tapered fixed-point).

Format	Keyword Spotting				Image classification			
	8-bit	7-bit	6-bit	5-bit	8-bit	7-bit	6-bit	5-bit
P	91.97%	85.33%	48.62%	23.79%	85.31%	77.28%	53.78%	26.19%
FP	86.45%	46.29%	13.72%	8.50%	82.10%	69.45%	11.28%	10.79%
FX	12.70%	8.87%	8.39%	8.22%	37.00%	21.80%	15.97%	12.90%
GP	**92.10%**	**91.39%**	**88.14%**	**49.62%**	**85.81%**	**83.90%**	**76.36%**	**42.72%**
TFX	79.20%	43.75%	8.52%	8.43%	85.24%	82.10%	38.60%	17.72%
32-bit FP	92.15%				86.26%			

Format	Visual Wake Words			
	8-bit	7-bit	6-bit	5-bit
P	83.02%	80.58%	74.53%	66.28%
FP	80.02%	68.25%	59.95%	59.37%
FX	76.43%	72.06%	61.86%	60.71%
GP	**83.02%**	**82.14%**	**76.72%**	**69.97%**
TFX	82.97%	81.92%	76.00%	66.76%
32-bit FP	82.26%			

discrepancy between the dynamic range provided by these numerical formats and the actual dynamic range of weights and activations.

5.2 ACTION Framework Results

Figures 4 and 5 illustrate the performance of each numerical format incorporated on the various configurations of accelerator and dataflows. The ILP optimization identified the optimal numerical format much quicker (≤ 1 s, performed on Intel i9-9960X) than the tedious and iterative process undertaken by reinforcement learning optimization algorithms (which can take several hours [5]). When constraints are selected in the region beyond the mean plus standard deviation of metrics (highlighted region), generalized posit was most frequently selected as the optimal numerical format. Note that except the accuracy vs. MAC frequency trade-off (Figs. 4.b and 5.b), the numerical formats are selected in way that to maximize accuracy when the accuracy and hardware constraints (e.g., EDP) are met. In the case of MAC frequency, the numerical formats are selected to maximize frequency when the accuracy constraints are satisfied.

To evaluate the efficacy of the numerical formats on a custom accelerator, the framework uses the SCALE-Sim simulator which outputs the estimated cycle

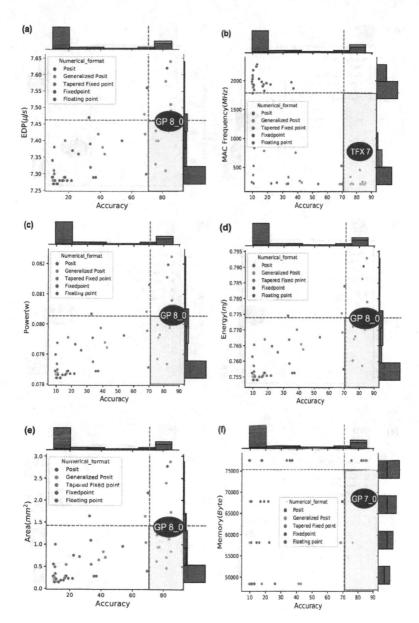

Fig. 4. (a) EDP vs Accuracy (b) MAC frequency vs. Accuracy (c) Power vs. Accuracy (d) Energy vs. Accuracy (e) Area vs Accuracy (f) Memory vs Accuracy for an image classification task with an accelerator configured with PEs arranged in a 16 × 16 systolic array and output stationary dataflow. The constraint was derived by adding the mean with the standard deviation of the metric. The numerical format selected by the ILP optimizer (marked by the large dark blue oval) in the highlighted region identifies the format for which the best accuracy and metric combination is achieved. GP n_es is n-bit generalized posit with es-bit exponent.

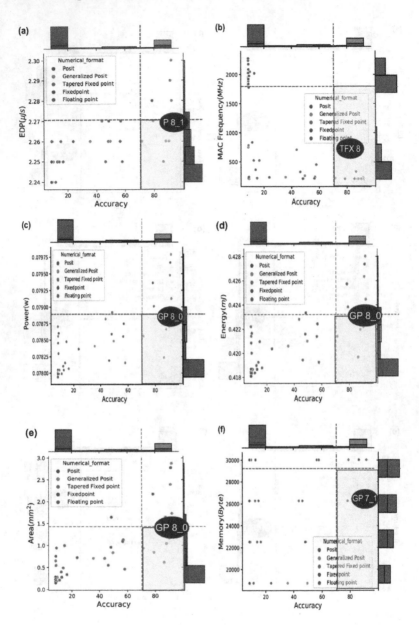

Fig. 5. (a) EDP vs Accuracy (b) MAC frequency vs. Accuracy (c) Power vs. Accuracy (d) Energy vs. Accuracy (e) Area vs Accuracy (f) Memory vs Accuracy for keyboard spotting task with an accelerator configured with PEs arranged in a 16 × 16 systolic array and output stationary dataflow. The constraint was derived by adding the mean with the standard deviation of the metric. The numerical format selected by the ILP optimizer (marked by the large dark blue oval) in the highlighted region identifies the format for which the best accuracy and metric combination is achieved. GP n_es is n-bit generalized posit with es-bit exponent.

count and the memory data movement sequences for executing a CNN model on a custom configuration. The cycle count and the data movement are combined with results obtained by synthesizing the MAC unit of each numerical format to generate the EDP and latency of the accelerator. Various dataflows and PE matrix array configurations were compared against the output stationary dataflow, which outperformed the other dataflows and offered a 24% reduction in latency as compared to the weight stationary dataflow in particular, for a single inference cycle. It has also shown significant improvement in EDP and utilization factor as compared to the input stationary and weight stationary dataflows. Moreover, generalized posit has outperformed all the other formats in Keyword Spotting and Image Classification tasks with minor overhead in EDP.

6 Conclusions

Through the ACTION framework, we propose a hardware-software co-design framework for early stage design space exploration to discover the optimal numerical formats and accelerator configurations based on custom user defined constraints. The configuration selection problem is solved by integer linear programming (ILP), which allows us to identify the optimal numerical format and accelerator configuration faster than reinforcement learning approaches. We show that generalized posit yields a 16% improvement in the average inference accuracy over the other numerical formats that are considered in this paper.

Acknowledgement. This research was supported by the Matrix AI Consortium for Human Well-Being at UTSA. The authors would like to thank Dr. John Gustafson, who is the inventor of Posit, Generalized posit and Tapered Fixed-point and has provided valuable insights over the years. The authors would also like to express gratitude to NUAI lab members at RIT and UTSA who supported this research study.

References

1. Banbury, C., et al.: Mlperf tiny benchmark. In: Proceedings of the Neural Information Processing Systems Track on Datasets and Benchmarks (2021)
2. Curnick, D.J., et al.: Smallsats: a new technological frontier in ecology and conservation? Remote Sens. Ecol. Conserv. 8(2), 139–150 (2021)
3. Reddi, V.J., et al.: Widening access to applied machine learning with tinyml. arXiv preprint arXiv:2106.04008 (2021)
4. Banbury, C.R., et al.: Benchmarking tinyml systems: Challenges and direction. arXiv preprint arXiv:2003.04821 (2020)
5. Wang, K., Liu, Z., Lin, Y., Lin, J., Han, S.: Haq: hardware-aware automated quantization with mixed precision. In: Proceedings of the IEEE/CVF Conference on Computer Vision and Pattern Recognition, pp. 8612–8620 (2019)
6. Rusci, M., Fariselli, M., Capotondi, A., Benini, L.: Leveraging automated mixed-low-precision quantization for tiny edge microcontrollers. In: Gama, J., et al. (eds.) ITEM/IoT Streams -2020. CCIS, vol. 1325, pp. 296–308. Springer, Cham (2020). https://doi.org/10.1007/978-3-030-66770-2_22

7. Langroudi, H.F., Karia, V., Pandit, T., Kudithipudi, D.: Tent: Efficient quantization of neural networks on the tiny edge with tapered fixed point. arXiv preprint arXiv:2104.02233 (2021)
8. Ghamari, S., et al.: Quantization-guided training for compact tinyml models. arXiv preprint arXiv:2103.06231 (2021)
9. Fahim, F., et al.: hls4ml: an open-source codesign workflow to empower scientific low-power machine learning devices. arXiv preprint arXiv:2103.05579 (2021)
10. Ravaglia, L., Rusci, M., Nadalini, D., Capotondi, A., Conti, F., Benini, L.: A tinyml platform for on-device continual learning with quantized latent replays. IEEE J. Emerg. Sel. Top. Circuits Syst. **11**(4), 789–802 (2021)
11. Lin, J., Chen, W.M., Cai, H., Gan, C., Han, S.: Mcunetv2: memory-efficient patch-based inference for tiny deep learning. arXiv e-prints pp. arXiv-2110 (2021)
12. Langroudi, H.F., Carmichael, Z., Pastuch, D., Kudithipudi, D.: Cheetah: mixed low-precision hardware & software co-design framework for dnns on the edge. arXiv preprint arXiv:1908.02386 (2019)
13. Hashemi, S., Anthony, N., Tann, H., Bahar, R.I., Reda, S.: Understanding the impact of precision quantization on the accuracy and energy of neural networks. In: Design, Automation & Test in Europe Conference & Exhibition (DATE), pp. 1474–1479. IEEE (2017)
14. Gustafson, J.L., Yonemoto, I.T.: Beating floating point at its own game: posit arithmetic. Supercomputing Front. Innov. **4**(2), 71–86 (2017)
15. Brumar, I., Zacharopoulos, G., Yao, Y., Rama, S., Wei, G.Y., Brooks, D.: Early DSE and automatic generation of coarse grained merged accelerators. arXiv preprint arXiv:2111.09222 (2021)
16. Yao, Z., et al.: Hawq-v3: Dyadic neural network quantization. In: International Conference on Machine Learning, pp. 11875–11886. PMLR (2021)
17. Hubara, I., Nahshan, Y., Hanani, Y., Banner, R., Soudry, D.: Improving post training neural quantization: layer-wise calibration and integer programming. arXiv preprint arXiv:2006.10518 (2020)
18. Gustafson, L.J.: A generalized framework for matching arithmetic format to application requirements. https://posithub.org/ (2020)
19. Langroudi, H.F., et al.: Alps: adaptive quantization of deep neural networks with generalized posits. In: Proceedings of the IEEE/CVF Conference on Computer Vision and Pattern Recognition, pp. 3100–3109 (2021)
20. Gysel, P., Pimentel, J., Motamedi, M., Ghiasi, S.: Ristretto: a framework for empirical study of resource-efficient inference in convolutional neural networks. IEEE Trans. Neural Netw. Learn. Syst. **29**(11), 5784–5789 (2018)
21. Langroudi, H.F., Karia, V., Gustafson, J.L., Kudithipudi, D.: Adaptive posit: parameter aware numerical format for deep learning inference on the edge. In: Proceedings of the IEEE/CVF Conference on Computer Vision and Pattern Recognition Workshops, pp. 726–727 (2020)
22. Tambe, T., et al.: Algorithm-hardware co-design of adaptive floating-point encodings for resilient deep learning inference. In: 2020 57th ACM/IEEE Design Automation Conference (DAC), pp. 1–6. IEEE (2020)
23. Samajdar, A., Zhu, Y., Whatmough, P., Mattina, M., Krishna, T.: Scale-sim: Systolic CNN accelerator simulator. arXiv preprint arXiv:1811.02883 (2018)
24. Abadi, M., et al.: TensorFlow: large-scale machine learning on heterogeneous systems (2015). https://www.tensorflow.org/, software available from tensorflow.org
25. Li, S., Romaszkan, W., Graening, A., Gupta, P.: Swis-shared weight bit sparsity for efficient neural network acceleration. arXiv preprint arXiv:2103.01308 (2021)

26. Zhou, C., et al.: Analognets: Ml-hw co-design of noise-robust tinyml models and always-on analog compute-in-memory accelerator. arXiv preprint arXiv:2111.06503 (2021)
27. Gousev, E.: Recent progress on tinyml technologies and opportunities. https://www.youtube.com/ (2021)
28. Banbury, C., et al.: Micronets: neural network architectures for deploying tinyml applications on commodity microcontrollers. In: Proceedings of Machine Learning and Systems, vol. 3 (2021)

MULTIPOSITS: Universal Coding of $\mathbb{R}^n$

Peter Lindstrom(✉)

Lawrence Livermore National Laboratory, Livermore, CA 94550, USA
`pl@llnl.gov`

Abstract. Recently proposed real-number representations like POSITS and ELIAS codes provide attractive alternatives to IEEE floating point for representing real numbers in science and engineering applications. Many of these applications represent fields on structured grids that exhibit smoothness, where adjacent scalar values are similar and often accessed together in stencil or vector computations. This similarity results in redundancy in representation, where several leading bits in the representation of adjacent values are shared.

We propose a generalization of scalar "universal codes" to small, multidimensional blocks of values that exploit their similarity and underlying dimensionality. Drawing upon ideas from multimedia and floating-point compression, our approach combines a decorrelating transform with adaptive, error-optimal interleaving of coefficient bits, which allows increasing accuracy per bit stored by orders of magnitude. Our solution accommodates both a fixed-length representation of blocks—facilitating random access—and variable-length storage to within a user-prescribed tolerance—e.g., for I/O, communication, and streaming computations. Our approach generalizes universal coding of the reals to vectors and tensors, and is straightforward to implement for several known number systems by extending a previously published framework for universal coding based on simple refinement rules.

Keywords: Number representations · Floating point · Universal coding · Data compression · Decorrelating transform · Vector quantization

1 Introduction

As data movement and storage have come to dominate the power and performance landscape in high-performance computing, there has been a recent push to investigate new real number representations that are more economical than the ubiquitous IEEE 754 floating-point format [1]. Example proposals include BRAIN-FLOATS [11] and TENSORFLOATS [2], which make a different tradeoff between the number of exponent and significand bits than IEEE 754. More significant departures from IEEE 754 include include UNUMS [8], POSITS [9], URRS [10], and ELIAS codes [20], some of which generalize universal codes originally developed for positive integers [7] to the reals. Many of these representations can be synthesized using number system frameworks that allow experimenting with alternative

© The Author(s), under exclusive license to Springer Nature Switzerland AG 2022
J. Gustafson and V. Dimitrov (Eds.): CoNGA 2022, LNCS 13253, pp. 66–83, 2022.
https://doi.org/10.1007/978-3-031-09779-9_5

number representations [18–20, 23, 25], and several examples have demonstrated the benefits of such representations in numerical applications in terms of improving the accuracy per bit stored [14, 16, 20]. Common to the universal number representations is the notion of *tapered accuracy* [21], where commonly occurring numbers near one are represented more accurately than rare numbers that are extremely small or large in magnitude. This is achieved by allocating fewer bits to represent the exponent in favor of retaining more bits for the significand (or fraction) for numbers near ±1.

Many science and engineering applications model the physical world as mostly continuous scalar fields, such as temperature and pressure, that are sampled discretely onto uniform Cartesian grids and are represented as multidimensional arrays of reals. In these applications, values at adjacent grid points tend to exhibit significant correlation, which manifests as shared leading bit patterns in their number representation. The conventional approach of representing arrays as independent scalars wastes precious bits on such redundant information, resulting in a larger than necessary memory footprint and associated costs in moving data through the memory hierarchy. Recent efforts have attempted to remove the redundancy using variations on block-floating-point representation [12], by partitioning arrays into small blocks of correlated scalars and eliminating shared information [3, 16, 17]. Whether explicit or not, such methods substitute the *scalar quantization* of reals implied by the number representation with a *vector quantization* step, where each fixed-length codeword encodes a whole block of numbers (unraveled as a vector). Current block-floating-point representations are modeled on IEEE 754—they use a fixed-length exponent common to the block and a set of significands (or coefficients) that are scaled by the common exponent.

In this paper, we propose an alternative block-based representation that builds upon the ideas shared by universal number representations, which use a variable-length encoding of the exponent and that—given sufficient precision—can represent *any* real. This is unlike IEEE 754 and block-floating-point representations that utilize a fixed-length exponent, which places a fixed limit on the smallest and largest numbers representable regardless of precision. Our new representation further reduces redundancy by performing a decorrelating linear transformation, which in effect replaces leading bits shared among values in a block with strings of leading zeros (or ones) that can be efficiently encoded. We demonstrate the accuracy benefits of a tapered number system for blocks of reals combined with a decorrelation step that eliminates shared leading bits in order to represent more trailing bits of significands. Like most other floating-point-like representations, we may truncate the binary representation at any point—a step analogous to *rounding*—to achieve a fixed-length representation of each block that facilitates random access (at block granularity). We may alternatively truncate the representation when it satisfies an error tolerance, resulting in variable-length records. In applications where the data is accessed sequentially, e.g., in I/O and streaming computations, such variable-length codes ensure a uniform level of error and avoid an excess in precision when it is not needed or when the trailing bits are already contaminated with error, e.g., due to roundoff, discretization, approximate solvers, sensor noise, etc. [6,15,28].

Our framework generalizes universal codes for $\mathbb{R}$ to $\mathbb{R}^n$ without being tied to any particular number representation. Rather, we allow any universal code for the reals to be used and show how to optimally interleave the bits representing a collection of scalars from such a code to represent decorrelated real-valued vectors or tensors. In this paper, we present results of extending POSITS and demonstrate their utility in multiple applications.

2   Preliminaries

One of our earlier insights [18] is that most real number representations are fully described by a cumulative distribution function (CDF), $F(x)$, with associated probability density, $f(x)$. $F(x)$ maps a real, $x \in \mathbb{R}$, to the interval $(0, 1)$, with $F(-\infty) = 0$ and $F(+\infty) = 1$.[1] The binary bit string $0.b_1 b_2 \ldots b_p$ thus represents $F(x) = \sum_{i=1}^{p} b_i 2^{-i} \in [0, 1)$ using p bits of precision.[2] In other words, for finite p, $F(x)$ is rounded to the nearest multiple of 2^{-p}.[3] This rounding may also be viewed as linear scalar quantization with step size $\Delta = 2^{-p}$.

A *universal code* for the reals also satisfies the following properties:

1. $f(x) > 0 \; \forall x \in \mathbb{R}$, which ensures that every real can be represented uniquely. Because $f(x) = 0$ for $x > \texttt{FLT\_MAX}$, IEEE 754 is not a universal code.
2. $|x| \le |y| \iff f(x) \ge f(y)$. In other words, $f(x)$ decreases monotonically away from zero, with larger $|x|$ requiring longer codewords. As a corollary, $f(x) = f(-x)$.
3. $\lim_{x \to \infty} \frac{-\log_2 f(x)}{\log_2 x}$ is finite. This ensures that $-\log_2 f(x)$, which governs code length, does not increase too rapidly. Like universal codes for integers, this property disqualifies representations like the unary code, whose length is arbitrarily longer than binary code.

These properties essentially generalize similar properties required for universal integer codes; see [7].

Another key insight from [18] is that universal codes may be expressed as two functions: a generator function, g, that is used in unbounded search to bracket $x \ge 1$ or $x^{-1} \ge 1$, and a refinement function, $r(x_{\min}, x_{\max})$, that is used in binary search to increase the precision by narrowing the interval containing x. For POSITS, $g(x) = \beta x$, where $\beta = 2^{2^m}$ is the *base* (also called *useed* in [9]); see [19]. In the 2022 POSIT standard, $m = 2$ for POSITS regardless of precision; thus, $g(x) = 16x$. The generalization of the ELIAS gamma code uses

---

[1] In POSITS and [20], $-\infty$ and $+\infty$ map to the same point, called NaR, and are represented as $F(-\infty) \bmod 1 = F(+\infty) \bmod 1 = 0$.

[2] Using two's complement representation, it is common to translate $(0, 1)$ to $(-\frac{1}{2}, \frac{1}{2})$ by negating the $b_1 2^{-1}$ term such that the bit string $0.000\ldots$ represents $x = 0$.

[3] Special rounding modes may be used so that finite numbers are not rounded to the interval endpoints $\{0, 1\}$, which represent $\pm\infty$.

$g(x) = 2x$, and is thus a particular instance of POSITS with $m = 0$, while ELIAS delta uses $g(x) = 2x^2$; see [18]. These number systems use as refinement function the arithmetic mean of the interval endpoints, $r(a, b) = (a + b)/2$, unless a and b differ in magnitude by more than 2, in which case the function returns the geometric mean, $r(a, b) = \sqrt{ab}$. Thus, a whole number system may be designed by specifying two usually very simple functions. For simplicity of notation, we will use the refinement function, r, to split intervals even when they are unbounded. We use the convention $r(-\infty, \infty) = 0$, $r(0, \infty) = 1$, $r(a, \infty) = g(a)$, $r(0, b) = r(b^{-1}, \infty)^{-1}$, $r(-b, -a) = -r(a, b)$; see [18]. Each codeword bit thus determines in which subinterval, $[a, s)$ or $[s, b)$, x lies, with $s = r(a, b)$.

A naïve way of extending universal codes from $\mathbb{R}$ to $\mathbb{R}^n$ would be to interleave the bits from the n independent codewords in round-robin fashion. Not only does this often result in significant redundancy of identical bits, but such round-robin interleaving is also suboptimal from an accuracy standpoint when absolute rather than relative error matters. As we shall see later, one can significantly improve accuracy by instead interleaving the bits in a data-dependent order.

3   Universal Coding of Vectors

Before describing our coding algorithm in detail, we first outline its key steps. Given a d-dimensional scalar array, we first partition it into blocks. As in the ZFP [17] representation, we have chosen our blocks to be of length 4 in each dimension. This block size has proven large enough to expose sufficient correlation, yet small enough to provide sufficiently fine granularity and allow a fast and simple implementation of decorrelation. Furthermore, a block size that is a power of two simplifies indexing via bitwise shifting and masking instead of requiring integer division. Hence, a d-dimensional block consists of $n = 4^d$ scalars. If array dimensions are not multiples of four, we pad blocks as necessary; see [6].

Each block is then encoded separately, either using a fixed number of bits, np, where p is the per-scalar precision, or by emitting only as many bits as needed to satisfy an absolute error tolerance, ϵ. The encoding step begins by decorrelating the block (Sect. 3.1) using a linear transformation. The goal of this step is to eliminate any correlation between values and to *sparsify* the block such that common leading bits in a scalar representation of the values are (usually) replaced with leading zero-bits, with many transform coefficients having small magnitude. We use the same transformation as in ZFP, which can be implemented very efficiently using addition, subtraction, and multiplication by $\frac{1}{2}$ or by 2. As in ZFP, the resulting set of n transform coefficients is then reordered by *total sequency* using a fixed permutation $\pi(x)$ (Sect. 3.2).[4] Finally, the reordered coefficients are encoded by emitting one bit at a time from the universal scalar code of one of the n coefficients selected in each iteration (Sect. 3.3). We leave

---

[4] Sequency denotes the number of zero-crossings of a discrete 1D function. Total sequency denotes the sum of per-component zero-crossings of basis functions in a tensor product basis, and is analogous to total degree of a multivariate polynomial.

$\text{ENCODE}(x = (x_1, \ldots, x_n), p, \epsilon)$
1. $x \leftarrow Qx$ // §3.1: decorrelate x
2. $x \leftarrow \pi(x)$ // §3.2: sort x on sequency
3. $R \leftarrow (-\infty, +\infty)^n$ // initialize hyper-rectangle
4. **for** $k = 1, \ldots, np$ // §3.3: encode x in up to np bits
5. $i \leftarrow \text{argmax}_j |R_j|$ // identify widest dimension i
6. $I \leftarrow R_i$
7.   **if** $|I| \leq \left(\frac{4}{15}\right)^d \epsilon$ // is error tolerance met?
8.   **terminate**
9. $s \leftarrow r(I_{\min}, I_{\max})$ // compute interval split point
10.   **if** $x_i < s$ // is x_i below the split point s?
11.   output(0)
12. $I_{\max} \leftarrow s$
13.   **else**
14.   output(1)
15. $I_{\min} \leftarrow s$
16. $R_i \leftarrow I$ // update narrowed interval

Listing 1: Universal coding algorithm for n-dimensional ($n = 4^d$) vector x using precision, p, and error tolerance, ϵ. If unspecified, we assume $\epsilon = 0$. $|I|$ denotes the interval width $I_{\max} - I_{\min}$.

the choice of universal scalar code open; in our algorithm, this choice is determined solely by the interval refinement function, r. The coefficient chosen in each iteration of the coding algorithm is data-dependent and done in a manner so as to minimize the upper bound on the L_∞ error in the reconstructed block. The decoding algorithm performs the same sequence of steps but using their natural inverses and in reverse order.

Listing 1 gives pseudocode for encoding a single d-dimensional block of $n = 4^d$ real scalars. Any arithmetic performed and state variables used must have sufficient precision, e.g., IEEE double precision. We proceed by describing each step of the encoding algorithm in more detail and conclude with some implementation details.

3.1 Decorrelation

When a (mostly) continuous function is sampled onto a sufficiently fine uniform grid, values at adjacent grid points tend to be significantly correlated. Such discrete data is said to be "smooth" or to exhibit autocorrelation. Autocorrelated data is undesirable because it introduces overhead in the representation, as exponents and leading significand bits of adjacent values tend to agree. The process of removing correlation is called *decorrelation*, which can be achieved using a linear transformation (i.e., by a matrix-vector product).

Consider a partitioning of a d-dimensional array into equal-sized blocks of $n = 4^d$ values each. Then each of the n "positions" within a block may be

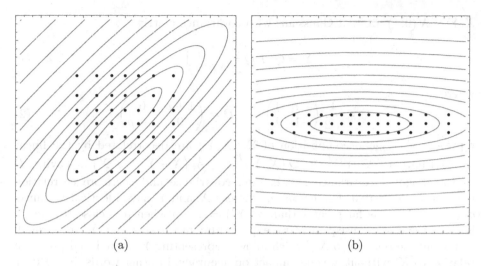

(a)   (b)

Fig. 1. Decorrelation of two correlated and identically distributed variables (X, Y) with variance $\sigma^2 = 1$ and covariance $\rho\sigma^2 = 0.9$. In this simple example, the decorrelating transform is merely a 45-degree rotation. These plots show contours of the joint probability density before and after decorrelation as well as the set of representable vectors (indicated by dots at regular quantiles) using 6 total bits of precision. In (a), vectors are represented using 3 bits per component. In (b), 4 bits are used for the X component (with variance $1 + \rho = 1.9$), while 2 bits are used for the Y component (with variance $1 - \rho = 0.1$). Notice the denser sampling and improved fit to the density in (b). Decorrelation here removes the covariance between the two variables.

associated with a random variable, X_i, with the values from the many blocks constituting random variates from the n random variables. The spatial correlation among the $\{X_i\}$ is determined by their variance and covariance. The covariance—and therefore correlation—is eliminated by performing a transformation (or change of basis) using a particular orthogonal $n \times n$ matrix, Q. For perfect decorrelation, this matrix Q is given by the eigenvectors of the covariance matrix, and the associated optimal transform is called the Karhunen-Loève Transform (KLT) [27]. The KLT is data-dependent and requires a complete analysis of the data, which is impractical in applications where the data evolves over time, as in PDE solvers. Instead, it is common to use a fixed transform such as the discrete cosine transform (DCT) employed in JPEG image compression [26], the Walsh-Hadamard transform [27], or the Gram orthogonal polynomial basis, which is the foundation for the transform used in the current version of ZFP [6] as well as in our encoding scheme. Such suboptimal transforms do not entirely eliminate correlation, though in practice they tend to be very effective.

To visualize the process of decorrelation, Fig. 1 shows a cartoon illustration using two correlated and identically distributed random variables (X, Y), representing the relationship between pairs of adjacent grid points. In Fig. 1(a),

$X, Y \sim \mathcal{N}(0, 1)$ are unit Gaussians with covariance matrix

$$\Sigma = Q^T DQ = \begin{pmatrix} 1 & \rho \\ \rho & 1 \end{pmatrix}, \tag{1}$$

where

$$Q = \frac{1}{\sqrt{2}} \begin{pmatrix} 1 & 1 \\ -1 & 1 \end{pmatrix} \qquad D = \begin{pmatrix} 1+\rho & 0 \\ 0 & 1-\rho \end{pmatrix} \tag{2}$$

represent the eigendecomposition of Σ. (X, Y) are decorrelated by the linear transformation $(X' \ Y')^T = Q (X \ Y)^T = \frac{1}{\sqrt{2}} (X + Y \ Y - X)^T$, leaving D as the diagonal covariance matrix, i.e., covariance and therefore correlation have been eliminated. Furthermore, whereas X and Y have identical variance, $\sigma^2(X') = 1 + \rho$ is far greater than $\sigma^2(Y') = 1 - \rho$ when ρ is close to one, as is often the case. Consequently, we expect random variates from Y' to be small in magnitude relative to X', which allows representing Y' at reduced precision relative to X' without adverse impact on accuracy. In other words, in a POSIT or other universal coding scheme of (X', Y') as independent components, with $|X'| \gg |Y'|$, the leading significand bits of X' carry more importance than the leading significand bits of Y', which have smaller place value.

Our approach to universal encoding of $\mathbb{R}^n$ is to make use of the independent scalar universal codes of the vector components (e.g., X' and Y' above), but to interleave bits from those codes by order of importance, i.e., by impact on error. This allows for fixed-precision representation of vectors from $\mathbb{R}^n$ (in our example, $n = 2$) as a fixed-length prefix of the full-precision bit string of concatenated bits. We may also use a variable-precision representation, where we keep all bits up to some minimum place value $\epsilon = 2^e$, where ϵ represents an absolute error tolerance.

Decorrelating Transform. The example above shows a decorrelating transform for pairs of values. In practice, in numerical applications where physical fields are represented (e.g., temperature on a 3D grid), values vary slowly and smoothly, and correlations extend beyond just immediate neighbors. This observation is the basis for block compression schemes such as JPEG image compression [26] and ZFP floating-point compression [17], where larger d-dimensional blocks of values are decorrelated together, e.g., 8×8 in JPEG and $4 \times 4 \times 4$ in 3D ZFP. Due to its success in science applications, we chose to base our universal encoding scheme on the ZFP framework, which relies on a fast transform that approximates the discrete cosine transform used in JPEG:

$$Q = \frac{1}{16} \begin{pmatrix} 4 & 4 & 4 & 4 \\ 5 & 1 & -1 & -5 \\ -4 & 4 & 4 & -4 \\ -2 & 6 & -6 & 2 \end{pmatrix} \qquad Q^{-1} = \frac{1}{4} \begin{pmatrix} 4 & 6 & -4 & 1 \\ 4 & 2 & 4 & 5 \\ 4 & -2 & 4 & -5 \\ 4 & -6 & -4 & 1 \end{pmatrix} \tag{3}$$

This transform, which is slightly non-orthogonal, can be implemented very efficiently in place using *lifting steps* [5] and involves only 5 additions, 5 subtractions, and 6 multiplications by $\frac{1}{2}$ or 2, compared to 12 additions and 16 multi-plications for a standard matrix-vector multiplication; see [6] for details. Unlike in ZFP, we perform arithmetic in floating point, e.g., using IEEE double precision or, if desired, MPFR arbitrary-precision arithmetic. Note how $\|Q\|_\infty = 1$, which ensures that there is no range expansion during application of Q. Conversely, $\|Q^{-1}\|_\infty = \frac{15}{4}$. Thus, rounding errors in the transform coefficients may expand by as much as $\left(\frac{15}{4}\right)^d$ in d dimensions, which is accounted for in Listing 1, line 7.

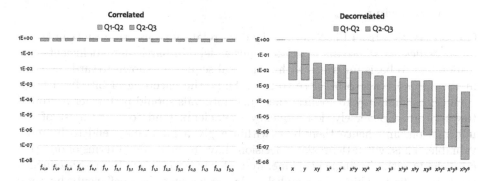

Fig. 2. Distributions as interquartile range of the magnitude relative to maximum for each of the 4×4 values $f_{i,j}$ (left) and transform coefficients for basis functions $O(x^i y^j)$ (right). The distributions represent four million randomly sampled 2D blocks from 32 fields from seven different data sources. Notice the effectiveness of decorrelation in compacting the signal energy into the low-order modes (vertical axis is logarithmic).

As in ZFP, we make use of a tensor product basis in $d > 1$ dimensions, where we apply the transform along each dimension of a block to decorrelate its 4^d values. Following decorrelation of a block, we proceed by encoding its transform coefficients, most of which tend to be very small in magnitude. Each block is thus transformed and encoded independent of other blocks, allowing access to arrays at block granularity.

Figure 2 illustrates the benefits of decorrelation by plotting the distributions of values from 2D blocks before and after decorrelation. These distributions represent the magnitude of values in each block relative to the block's largest value, i.e., $|f_{i,j}| / \max_{0 \le i,j \le 3} |f_{i,j}|$. The plots show how essentially identically distributed values $f_{i,j}$ are sparsified by decorrelation using the ZFP tensor product basis $Q \otimes Q$. Each basis vector approximates a regularly sampled orthogonal Gram polynomial, with coefficients for high-order polynomial terms being several orders of magnitude smaller than the constant and linear terms. This implies that the data within each block is well approximated using only a few low-order terms. The basis functions $O(x^i y^j)$ have been ordered by total degree $i + j$, then by $i^2 + j^2$, resulting in a nearly monotonic decrease in each of the quartiles ($Q1 = 25\%$, $Q2 = 50\% = $ median, $Q3 = 75\%$).

Using a Taylor expansion of the continuous scalar field being encoded, one can show that the magnitude of the ith transform coefficient, f_i', in $d = 1$

dimension varies as $O(h^i)$, with h being the grid spacing. The extension to higher dimensions is straightforward, e.g., $|f'_{i,j,k}| = O((\Delta x)^i (\Delta y)^j (\Delta z)^k)$. Thus, $|f'_{i,j,k}| = O(h^{i+j+k})$ when $\Delta x = \Delta y = \Delta z = h$. This further explains why total sequency ordering by $i + j + k$ results in a monotonic decrease in magnitude as $h \to 0$. Moreover, this has implications on the variance of transform coefficients, which could be exploited if different scalar coding schemes were used for the transform coefficients.

3.2 Reordering

One observation from Fig. 2 is that the decorrelating transform results in transform coefficients whose distributions differ widely. In particular, coefficients corresponding to basis functions with high total sequency (shown toward the right in this figure) tend to be close to zero. Thus, the encoding of bits from those coefficients tend only to confirm that their intervals should be further narrowed toward zero. Conversely, low-sequency coefficients tend to carry most of the information, and hence their bits (within a given bit plane) tend to be more valuable. Thus, when tiebreaking decisions have to be made in terms of ordering coefficients within a single bit plane, we prefer to encode bits from low-sequency coefficients first. This is accomplished by reordering the coefficients by total sequency, as is also done in ZFP and JPEG. That is, a coefficient $f'_{i,j,k}$ in a 3D block has total sequency $i + j + k$. We use as secondary sort key $i^2 + j^2 + k^2$, e.g., a trilinear term ($i = j = k = 1$) precedes a cubic one ($i = 3, j = k = 0$), and break any remaining ties arbitrarily. Note that this ordering tends to list coefficients roughly by decreasing magnitude.

3.3 Encoding

At this point, we have a set of decorrelated values roughly ordered by decreasing magnitude. Because they are no longer correlated, their joint probability density (in the idealized case) is given by the product of marginal densities:

$$f(X_1, X_2, \ldots, X_n) = f_1(X_1) f_2(X_2) \cdots f_n(X_n). \tag{4}$$

Due to this independence, vector quantization is reduced to independent scalar quantization, where the quantization results in an n-dimensional "grid" onto which the vector $X = (X_1, \ldots, X_n)$ is quantized. Note that such a grid need not have the same number of grid points (as implied by the per-variable precision) along each dimension.

Though the X_i are independent, note that they are not identically distributed, as evidenced by Fig. 2. Ideally, we would design a separate code optimal for each such distribution, however this brings several challenges:

- The actual distributions are data or application dependent. While some efforts have been made to optimize number systems for given data distributions [13], such approaches become impractical in computations like PDE solvers, where the distributions are not known a priori.

- Even if the data distributions were known, finding corresponding error minimizing codes is an open problem. Currently, L_2 optimal codes are known for only a few distributions, most notably the Laplace distribution [22].
- Assuming these two prior challenges can be addressed, the CDF for an error optimal code would likely not be expressible in closed form or would involve nontrivial math functions that would be prohibitively expensive to evaluate. For best performance, we prefer CDFs that are linear over binades.

Faced with these challenges, we take a different approach by making use of "general purpose" universal number representations like Posits and by optimizing the order in which bits from the X_i are interleaved to minimize the L_∞ error norm. While representations like Posits are parameterized (on "exponent size"), which would allow parameter selection tailored to each random variable X_i, we do not pursue such an approach here but believe it would be a fruitful avenue for future work.

Given a codeword c comprised of interleaved bits, c can be thought of as encoding the path taken when traversing a k-d tree that recursively partitions the n-dimensional space—a hyper-rectangle—in halves using a sequence of binary cuts, each along one of the n axes. To minimize the L_∞ error norm, we should always cut the hyper-rectangle containing x along the axis in which it is widest. Due to the expected monotonic and rapid decrease in magnitude of the x_i, this suggests that the hyper-rectangle is usually wider for small i than for large i, and that a few leading bits for x_i with large i are sufficient to determine that such coefficients are small and contribute little to the overall accuracy. Hence, many leading bits of the codeword will be allocated to $x_{0,\ldots,0}$—the mean value within a block—while the bits for small, high-frequency components are deferred until later since they have only small impact on accuracy.

Our encoding algorithm tracks the interval endpoints for each x_i. In each iteration, corresponding to the output of a single bit, it conceptually sorts the intervals by width. For each codeword bit, we split the widest interval; when there is a tie, we prefer x_i with low index (i.e., total sequency), i. The resulting scheme effectively reduces to *bit plane coding* (cf. [17,24]), where the n bits of a bit plane are encoded together before moving on to the next significant bit. The bracketing sequence associated with universal coding, however, quickly prunes many bits of a bit plane by marking whole groups of bits of a coefficient as zero.[5] Consequently, by simple bookkeeping (through tracking intervals), many bits of a bit plane are known to be zero and need not be coded explicitly.

The decoding step proceeds in reverse order and progressively narrows the n intervals based on the outcomes of single bit tests. The result of this process is a set of intervals that x is contained in. Our current approach is to simply use the lower interval bound along each dimension as representative. Other strategies, such as using the next split point or by rounding the input vector during encoding could also be used, though the latter is complicated by not knowing a priori the precision of each vector component, which is data-dependent.

---

[5] This marking is done in variable-radix coding [19] by testing whole digits of radix $\beta > 2$, e.g., four bits at a time are tested in Posits with $\beta = 2^4 = 16$.

3.4   Implementation

Although the algorithm in Listing 1 is straightforward, it involves an expensive step to repeatedly find the widest of $n = 4^d$ intervals (line 5). A linear search requires $O(n)$ time, which can be accelerated (especially for $d \geq 3$) to $O(\log n)$ time using a heap data structure. In each iteration, we operate on one of the n intervals and keep the remaining $n - 1 = 2^{2d} - 1$ intervals sorted in a heap, which conveniently is a perfect binary tree with $2d$ levels. Following the narrowing of an interval (line 16), we compare its width to the heap root's, and if still larger, we continue operating on the same interval in the next iteration. Otherwise, we swap the current interval with the heap root and sift it down (using $O(\log n)$ operations) until the heap property has been restored. For 3D data, we found the use of a heap to accelerate encoding by roughly 4×.

We note that the implementation of universal vector codes presented here has not been optimized for speed. The need to perform arithmetic on intervals and to process a single bit at a time clearly comes at a substantial expense. We see potential speedups by tracking interval widths in terms of integer exponents instead. We may also exploit faster scalar universal coding schemes developed, for example, for POSITS, which process multiple bits at a time. Furthermore, it may be possible to avoid data-dependent coding by exploiting expected relationships between coefficient magnitudes such that the order in which bits are interleaved may be fixed. Such performance optimizations are left as future work.

4   Results

We begin our evaluation by examining the rate-distortion tradeoff when encoding static floating-point outputs from scientific simulations. Although it is common to compare representations by plotting the signal-to-noise ratio (SNR) as a function of rate—the number of bits of storage per scalar value—we have chosen to represent the same information in terms of what we call the *accuracy gain* vs. rate. We define the accuracy gain, α, as

$$\alpha = \log_2 \frac{\sigma}{E} - R = \frac{1}{2} \log_2 \frac{\sum_i (x_i - \mu)^2}{\sum_i (x_i - \tilde{x}_i)^2} - R, \qquad (5)$$

where σ and μ are the standard deviation and mean of the original data, x_i is one of the original data values and $\tilde{x}_i$ is its approximation in a given finite-precision number system, E is the L_2 error (distortion), and R is the rate. Here the term $\log_2 \frac{\sigma}{E}$ provides a lower bound on the rate required to encode an (uncorrelated) i.i.d. Gaussian source within error E [4, §10.3.2], and effectively serves as a baseline against which R is measured. For correlated data, we expect $R \leq \log_2 \frac{\sigma}{E}$ for a number representation that exploits correlation, resulting in $\alpha \geq 0$. Conversely, because scalar representations like IEEE 754 and POSITS ignore such correlations, they yield $\alpha \leq 0$. We note that α is high when the error, E, and the rate, R, are low. For effective coding schemes, $\alpha(R)$ tends

to increase from zero at low rates—indicating that "compression" is achieved—
until a stable plateau is reached, when each additional bit encoded results in
a halving of the error—indicating that random, incompressible significand bits
have been reached. Ultimately, E either converges to zero ($\alpha \to \infty$) or to some
small nonzero value, e.g., due to roundoff errors, where additional precision is not
helpful ($\alpha \to -\infty$). The maximum α indicates the amount of redundant information that a representation is able to eliminate. In addition, α allows comparing
the efficiency of representations when both R and E differ by a nonnegligible
amount, which using R, E, or SNR $= 20 \log_{10} \frac{\sigma}{E}$ alone would be difficult.

4.1   Static Data

Figure 3 plots the accuracy gain (higher is better) for various representations
of two fields (density and viscosity) from a hydrodynamics simulation.[6] The
density field varies in the range $[1, 3]$ while the viscosity field spans many orders
of magnitude and also includes negative values.

The representations compared include IEEE 754 (half and float); POSITS and
ELIAS δ; two versions of MULTIPOSITS based on our universal vector coding
scheme; and two corresponding versions of ZFP. Here the -R suffix indicates
fixed-rate representations, where each block is assigned the same number of
bits; the -A suffix indicates fixed-accuracy representations, where a given error
tolerance dictates the storage size of each block. Fixed-accuracy mode is generally preferable when emphasis is on error rather than storage size, as then errors
are roughly uniform over the entire domain, which allows for a smaller storage
budget when the tolerance is met. In fixed-rate mode, additional bits are typically spent on each block, but the total L_2 error is usually dominated by the
highest-error blocks. Hence, reducing the error nonuniformly across blocks does
not appreciably reduce the total error but does increase storage. Of course, the
variable-rate storage associated with a fixed-accuracy representation complicates
memory management and random access, but we include such results here as
they serve an important use case: offline storage and sequential access.

The two plots in Fig. 3 suggest several trends. First, fixed-rate MULTIPOSITS
generally improve on POSITS by about 3.5–8 bits of accuracy across a wide range
of rates; POSITS in turn perform better than IEEE. The negative accuracy gain
for the scalar representations essentially corresponds to the overhead of encoding
exponents, and we see that IEEE does worse when using 8 (float) rather than 5
(half) exponent bits. In all cases, ZFP outperforms MULTIPOSITS, for reasons
that will be discussed below. We also see that fixing the accuracy (-A) rather
than rate (-R) is a substantial improvement. We note that this may be of importance for I/O and communication applications, where the data is serialized and
transferred sequentially. While we have implemented fixed-accuracy mode for
MULTIPOSITS, the same idea could be generalized to scalar representations like
POSITS and IEEE, i.e., by truncating any significand bits whose place value fall

---

[6] The double-precision fields are from the MIRANDA code and are available from SDR-
BENCH at https://sdrbench.github.io.

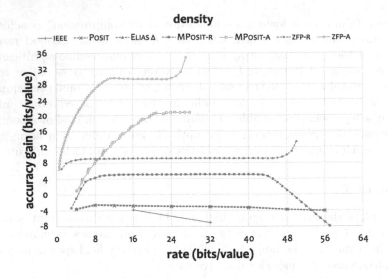

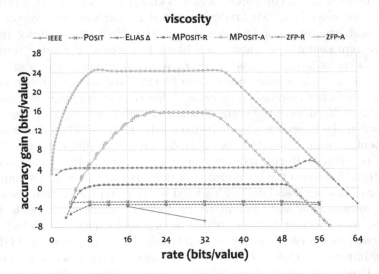

Fig. 3. Accuracy gain as a function of rate for two $384 \times 384 \times 256$ fields from a hydrodynamics simulation. The small dynamic range of the density field allows it be represented without loss using most representations, resulting in an eventual uptick in accuracy gain as the error approaches and even reaches zero. The viscosity field cannot be represented without loss, and the error eventually converges to a small value, resulting in an eventual decline in accuracy gain as additional bits do not reduce the error.

below some given power of two. Another observation is that MULTIPOSITS-A gives a somewhat irregular curve both in terms of rate and accuracy gain, in contrast to ZFP-A; there is a noticeable jump in both R and α every four data points at low rates. These jumps correlate with the increase in number of POSIT regime bits [9], which occur every time the binary exponent increases or decreases by four. Such increments introduce a flurry of additional bits for the next bit plane that increase both rate and accuracy. We finally note that some representations, like MULTIPOSITS, incur additional roundoff error from the use of double-precision arithmetic, as evidenced by the gap between MULTIPOSITS and ZFP at high rates.

4.2 Dynamic Data

We have implemented our universal vector code within the context of the ZFP framework, which accommodates user-defined codecs for its compressed-array C++ classes. These classes handle encoding, decoding, and caching of blocks (in IEEE double-precision format) for the user and expose a conventional multidimensional array API, thus hiding all the details of how the arrays are represented in memory. We additionally implemented a codec that uses a traditional scalar representation of blocks to allow for an apples-to-apples comparison using a single array implementation.

Based on these arrays, we implemented a 3D Poisson partial differential equation (PDE) solver using finite differences with Gauss-Seidel updates, i.e., array elements are updated in place as soon as possible. The equation solved is

$$\Delta u(x, y, z) = \sqrt{x^2 + y^2 + z^2} = r \qquad (6)$$

on $\Omega = [-1, 1]^3$ with boundary condition $u = \frac{1}{12}r^3$ and initial condition $u = 0$ on the interior of the domain. Given this setup, the closed form solution equals $u = \frac{1}{12}r^3$ on the entire domain. We use a standard second-order 7-point stencil for the Laplacian finite difference operator and a grid of dimensions 64^3. Higher-order stencils did not appreciably change the results.

Figure 4 plots the L_2 error in Δu as a function of solver iteration. As is evident, the low-precision scalar types quickly converge to a fixed error level as they run out of precision to accurately resolve differences. The MULTIPOSIT and ZFP vector types perform significantly better, both at 16- and 32-bit precision. Compared to IEEE 32-bit float, the 32-bit MULTIPOSIT representation improves the solution accuracy by five orders of magnitude.

5   Discussion

Our universal vector codes generalize the corresponding scalar codes for correlated multidimensional fields that often arise in scientific computing. Using a decorrelating step, we decouple the vector quantization step into independent scalar quantization steps and later interleave the bits from their binary representation so as to minimize error. Our framework relies on the simple and general framework from [18] to produce a codeword one bit at a time, which ensures a straightforward if inefficient implementation.

Our framework shares several steps with the ZFP number representation for multidimensional blocks:

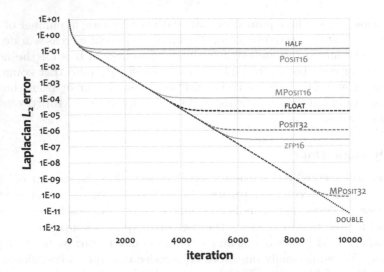

Fig. 4. Poisson equation solution error for various representations of the evolving state variable. Not shown is ZFP32, which coincides with the curve for DOUBLE.

- A decomposition of d-dimensional arrays into blocks of 4^d values.
- The same fast, linear, decorrelating transform. The key difference is that we implement our transform in floating point rather than integer arithmetic.
- The same reordering and prioritization of transform coefficients.

The two frameworks also differ in several ways:

- Whereas ZFP uses a fixed-length encoding of a single per-block exponent, we use per-coefficient tapered exponent coding.
- We inherit the same two's complement representation used for scalar POSITS, whereas ZFP encodes integer values in negabinary.
- ZFP makes use of additional control bits to encode the outcome of group tests, which apply to multiple bits within the same bit plane. Our framework does not use group testing across vector components but rather within each scalar to form regime bits. These bits directly refine the representation, like all significand bits, whereas ZFP's group tests instead govern the control flow.

In head-to-head competition, ZFP is a clear winner in terms of accuracy, storage, and speed, in part due to a more sophisticated coding scheme, though the speed advantage comes from its ability to process multiple bits simultaneously. Our framework as designed is data-dependent and operates at the single-bit level. ZFP also has the advantage of exploiting the sparsity of transform coefficients, which allows concise encoding of up to 4^d zeros in d dimensions using a single bit. In contrast, our scheme achieves only up to 4:1 "compression" of POSIT zero-bits and must encode at least 2×4^d bits to finitely bound each of the 4^d transform coefficients. By comparison, ZFP routinely allows a visually fair representation of 3D blocks using one bit per value or less. In fixed-rate mode, our framework like ZFP suffers from lack of proper rounding, as the per-coefficient precision is data-dependent and not known until encoding completes. Hence, coefficients are always rounded toward $-\infty$. We suggest possible strategies to combat the effects of improper rounding above. Finally, the tapered nature of POSITS and related number systems implies that blocks whose transform coefficients differ significantly from one may require many bits to even bracket the coefficients. In fact, the range preserving nature of the decorrelating transform on average causes already small coefficients to be reduced even further. ZFP performs some level of bracketing by aligning all values to a single common block exponent, which requires only a fraction of a bit per value to encode.

From these observations, we conclude that MULTIPOSITS offer a significant advantage over POSITS in applications that involve smooth fields while not rivaling the ZFP number system. Nevertheless, we believe that the ideas explored here may seed follow-on work to improve upon our framework, both with respect to accuracy per bit stored and speed. For instance, our coding scheme ignores the potential for intra bit plane compression and the potential to avoid data dependencies by adapting codes better suited to each of the transform coefficients.

6 Conclusion

We have presented a universal encoding scheme that generalizes the POSIT and other universal scalar number systems to vectors or blocks of numbers for numerical applications that involve spatially correlated fields. Our approach is to partition the data arrays into blocks, decorrelate the blocks using a fast transform, and then interleave bits from a universal coding of vector components in an error-optimal order. Using numerical experiments with real data and partial differential equation solvers, we demonstrated that MULTIPOSITS may yield as much as a six orders-of-magnitude increase in accuracy over conventional POSITS for the same storage, and even larger increases compared to IEEE 754 floating point. While our approach, as currently presented, is primarily of theoretical interest due to its high computational cost, we envision that our results will inspire follow-on work to address the performance issues associated with bitwise coding of vectors. In particular, we hope to develop data-independent universal vector codes that reap similar per-bit accuracy benefits with near-zero computational cost.

Acknowledgments. This work was performed under the auspices of the U.S. Department of Energy by Lawrence Livermore National Laboratory under Contract DE-AC52-07NA27344.

References

1. IEEE std 754-2019: IEEE standard for floating-point arithmetic (2019). https://doi.org/10.1109/IEEESTD.2019.8766229

2. Choquette, J., Gandhi, W., Giroux, O., Stam, N., Krashinsky, R.: NVIDIA A100 tensor core GPU: performance and innovation. IEEE Micro **41**(2), 29–35 (2021). https://doi.org/10.1109/MM.2021.3061394

3. Clark, M.A., Babich, R., Barros, K., Brower, R., Rebbi, C.: Solving lattice QCD systems of equations using mixed precision solvers on GPUs. Comput. Phys. Commun. **181**(9), 1517–1528 (2010). https://doi.org/10.1016/j.cpc.2010.05.002

4. Cover, T.M., Thomas, J.A.: Elements of Information Theory, 2nd edn. Wiley-Interscience (2005). https://doi.org/10.1002/047174882X

5. Daubechies, I., Sweldens, W.: Factoring wavelet transforms into lifting steps. J. Fourier Anal. Appl. **4**(3), 247–269 (1998). https://doi.org/10.1007/bf02476026

6. Diffenderfer, J., Fox, A., Hittinger, J., Sanders, G., Lindstrom, P.: Error analysis of ZFP compression for floating-point data. SIAM J. Sci. Comput. **41**(3), A1867–A1898 (2019). https://doi.org/10.1137/18M1168832

7. Elias, P.: Universal codeword sets and representations of the integers. IEEE Trans. Inf. Theory **21**(2), 194–203 (1975). https://doi.org/10.1109/tit.1975.1055349

8. Gustafson, J.L.: The End of Error: Unum Computing. Chapman and Hall (2015). https://doi.org/10.1201/9781315161532

9. Gustafson, J.L., Yonemoto, I.T.: Beating floating point at its own game: posit arithmetic. Supercomput. Frontiers Innov. **4**(2), 71–86 (2017). https://doi.org/10.14529/jsfi170206

10. Hamada, H.: URR: universal representation of real numbers. N. Gener. Comput. **1**, 205–209 (1983). https://doi.org/10.1007/bf03037427

11. Kalamkar, D., et al.: A study of BFLOAT16 for deep learning training (2019). https://doi.org/10.48550/arXiv.1905.12322

12. Kalliojarvi, K., Astola, J.: Roundoff errors in block-floating-point systems. IEEE Trans. Signal Process. **44**(4), 783–790 (1996). https://doi.org/10.1109/78.492531

13. Klöwer, M.: Sonum8 and Sonum16 with maximum-entropy training (2019). https://doi.org/10.5281/zenodo.3531887

14. Klöwer, M., Düben, P.D., Palmer, T.N.: Posits as an alternative to floats for weather and climate models. In: Conference for Next Generation Arithmetic, pp. 2.1–2.8 (2019). https://doi.org/10.1145/3316279.3316281

15. Klöwer, M., Razinger, M., Dominguez, J.J., Düben, P.D., Palmer, T.N.: Compressing atmospheric data into its real information content. Nat. Comput. Sci. **1**, 713–724 (2021). https://doi.org/10.1038/s43588-021-00156-2

16. Köster, U., et al.: Flexpoint: an adaptive numerical format for efficient training of deep neural networks. In: Conference on Neural Information Processing Systems, pp. 1740–1750 (2017). https://doi.org/10.48550/arXiv.1711.02213

17. Lindstrom, P.: Fixed-rate compressed floating-point arrays. IEEE Trans. Visual Comput. Graphics **20**(12), 2674–2683 (2014). https://doi.org/10.1109/TVCG.2014.2346458

18. Lindstrom, P.: Universal coding of the reals using bisection. In: Conference for Next Generation Arithmetic, pp. 7:1–7:10 (2019). https://doi.org/10.1145/3316279.3316286

19. Lindstrom, P.: Variable-radix coding of the reals. In: IEEE 27th Symposium on Computer Arithmetic (ARITH), pp. 111–116 (2020). https://doi.org/10.1109/ARITH48897.2020.00024

20. Lindstrom, P., Lloyd, S., Hittinger, J.: Universal coding of the reals: alternatives to IEEE floating point. In: Conference for Next Generation Arithmetic, pp. 5:1–5:14 (2018). https://doi.org/10.1145/3190339.3190344

21. Morris, R.: Tapered floating point: a new floating-point representation. IEEE Trans. Comput. **C-20**(12), 1578–1579 (1971). https://doi.org/10.1109/T-C.1971.223174

22. Noll, P., Zelinski, R.: Comments on "Quantizing characteristics for signals having Laplacian amplitude probability density function." IEEE Trans. Commun. **27**(8), 1259–1260 (1979). https://doi.org/10.1109/TCOM.1979.1094523

23. Omtzigt, E.T.L., Gottschling, P., Seligman, M., Zorn, W.: Universal numbers library: design and implementation of a high-performance reproducible number systems library (2020). https://doi.org/10.48550/arXiv.2012.11011

24. Said, A., Pearlman, W.A.: A new, fast, and efficient image codec based on set partitioning in hierarchical trees. IEEE Trans. Circuits Syst. Video Technol. **6**(3), 243–250 (1996). https://doi.org/10.1109/76.499834

25. Thien, D., Zorn, B., Panchekha, P., Tatlock, Z.: Toward multi-precision, multi-format numerics. In: IEEE/ACM 3rd International Workshop on Software Correctness for HPC Applications (Correctness), pp. 19–26 (2019). https://doi.org/10.1109/Correctness49594.2019.00008

26. Wallace, G.K.: The JPEG still picture compression standard. IEEE Trans. Consum. Electron. **38**(1), xviii–xxxiv (1992). https://doi.org/10.1109/30.125072

27. Wang, R.: Introduction to Orthogonal Transforms. Cambridge University Press (2012). https://doi.org/10.1017/cbo9781139015158

28. Wilkinson, J.: Error analysis of floating-point computation. Numer. Math. **2**, 319–340 (1960). https://doi.org/10.1007/BF01386233

Comparing Different Decodings for Posit Arithmetic

Raul Murillo[✉][iD], David Mallasén[iD], Alberto A. Del Barrio[iD],
and Guillermo Botella[iD]

Complutense University of Madrid, 28040 Madrid, Spain
{ramuri01,dmallase,abarriog,gbotella}@ucm.es

Abstract. Posit arithmetic has caught the attention of the research
community as one of the most promising alternatives to the IEEE 754
standard for floating-point arithmetic. However, the recentness of the
posit format makes its hardware less mature and thus more expensive than the floating-point hardware. Most approaches proposed so far
decode posit numbers in a similar manner as classical floats. Recently, a
novel decoding approach has been proposed, which in contrast with the
previous one, considers negative posits to have a negative fraction. In
this paper, we present a generic implementation for the latter and offer
comparisons of posit addition and multiplication units based on both
schemes. ASIC synthesis reveals that this alternative approach enables
a faster way to perform operations while reducing the area, power and
energy of the functional units. What is more, the proposed posit operators are shown to improve the state-of-the-art of implementations in
terms of area, power and energy consumption.

Keywords: Computer arithmetic · Posit · Decoding · Addition ·
Multiplication

1 Introduction

Historically, most scientific applications have been built on top of the IEEE 754
standard for floating-point arithmetic [10], which has been for decades the format
for representing real numbers in computers. Nevertheless, the IEEE 754 format
possesses some problems that are inherent to its construction, such as rounding,
reproducibility, the existence of signed zero, the denormalized numbers or the
wasted patterns for indicating *Not a Number* (NaN) exceptions [6]. All in all,
IEEE 754 is far from being perfect, as different CPUs may produce different
results, and all these special cases must be dynamically checked, which increases
the hardware cost of IEEE 754 units.

Recently, several computer arithmetic encodings and formats, such as the
High-Precision Anchored (HPA) numbers from ARM, the Hybrid 8-bit Floating
Point (HFP8) format from IBM, bfloat16, and many more have been considered
as an alternative to IEEE 754-2019 compliant arithmetic [7], which has also

© The Author(s), under exclusive license to Springer Nature Switzerland AG 2022
J. Gustafson and V. Dimitrov (Eds.): CoNGA 2022, LNCS 13253, pp. 84–99, 2022.
https://doi.org/10.1007/978-3-031-09779-9_6

recently included a 16-bit IEEE 754 version. Nonetheless, the appearance of the disruptive posit arithmetic [8] in 2017 has shaken the board. While the aforementioned approaches, except for the half-precision IEEE 754, are vendor-specific, posits aim to be standard. This novel way of representing reals mitigates and even solves the previously mentioned IEEE 754 drawbacks. Posits only possess one rounding mode, and there are just two special cases to check (zero and infinity). Also, posits are ordered in the real projective line, so comparisons are basically as the integer ones, and even conceive the use of fused operations in order to avoid losing precision. This is done by avoiding rounding of individual operations and accumulating the partial results in a large register called *quire*, which can even speed up computations with a large number of operands [15]. Another interesting property of posits is their tapered precision, that is, they are more accurate when their magnitude is in the proximity of zero, that is, their absolute value is near 1. These last properties have attracted a lot of attention from the community because they suit Deep Learning applications [3,9,12,17]. These applications leverage the multiply-accumulate (MAC) operations in order to accelerate the computation of matrix and dot products [2,23]. Furthermore, the numbers employed are typically normalized and thus fall in the proximities of zero. According to some authors, 32-bit posits can provide up to 4 orders of magnitude improvement in terms of accuracy [14,19] when comparing with the equivalent single-precision floating-point format. Nevertheless, this accuracy enhancement comes at a cost. The quire occupies a vast portion of the resulting posit functional unit [14,19,22].

Since 2017, several designs have appeared which implement individual [4,11,16] and fused [3,19,24] posit operators. While their implementations are different, either because of the functionality or due to the design, the unpacking/decoding of posits is common to all of them. This paper presents a study about the different ways of decoding posit numbers in literature, which directly affects how these decoding units unpack posit operands and that could also impact some other portions of the functional unit itself. Results show that decoding posits in a different manner to the classical one inspired by floating-point arithmetic can substantially reduce the hardware resources used by functional units. In addition, this work presents an implementation of posit functional units that follows the alternative decoding scheme aforementioned. The proposed implementation outperforms state-of-the-art designs of posit adders and multipliers in terms of performance and hardware requirements.

The rest of the paper is organized as follows: Section 2 introduces the necessary background about the posit format, and details the two different approaches for decoding posits that have been proposed so far. Section 3 describes the different components of fundamental posit arithmetic units (adders and multipliers), as well as the existing design differences when using each of the decoding approaches. The performance and resource utilization of both approaches are compared in Sect. 4, which shows ASIC synthesis results for different components and arithmetic units from the literature. Finally, Sect. 5 concludes this work.

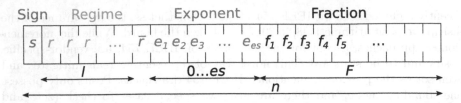

Fig. 1. Posit$\langle n, es \rangle$ binary encoding. The variable-length regime field may cause the exponent to be encoded with less than es bits, even with no bits if the regime is wide enough. The same occurs with the fraction.

2   Posit Arithmetic

A posit format is defined as a tuple $\langle n, es \rangle$, where n is the total bitwidth of the posits and es is the maximum number of bits reserved for the exponent field. As Fig. 1 shows, posit numbers are encoded with four fields: a sign bit (s), several bits that encode the regime value (k), up to es bits for the unsigned exponent (e), and the remaining bits for the unsigned fraction (f). The regime is a sequence of l identical bits r finished with a negated bit $\bar{r}$ that encodes an extra scaling factor. As this field does not have a fixed length, some exponent or fraction bits might not fit in the n-bit string, so 0 would be assigned to them. The variable length of this field allows posit arithmetic to have more fraction bits for values close to ± 1 (which increases the accuracy within that range), or to have less fraction bits for the sake of more exponent bits for values with large or small magnitudes (increasing this way the range of representable values). This is known as *tapered accuracy*, and contrasts with the constant accuracy that IEEE 754 floats present, due to the fixed length of the exponent and fraction fields, as can be seen in Fig. 2 (here, the left part of the IEEE floating-point format corresponds to the gradual underflow that subnormal numbers produce).

Posit arithmetic only considers two special cases: zero, that is represented with all bits equal to 0, and *Not a Real* (NaR) exception, represented by all the bits except the sign bit equal to 0. The rest of the bit patterns are used to represent a different real value. However, at the time of writing this paper, two main different ways of understanding how posit bit strings represent real values have been proposed: using the sign-magnitude format, as floating-point numbers, or considering posits in two's complement notation. While both approaches are equivalent from a mathematical sense (i.e. the same bit patterns represent the same values, regardless of the approach), they present implementation differences that should be considered when implementing such arithmetic format in a physical device. Other alternative interpretations of posits are discussed in [13].

2.1   Sign-Magnitude Posit Decoding

Posit arithmetic is a floating-point format for representing real numbers. Thus, the numerical value X of a normal posit datum was initially defined in [8] by (1)

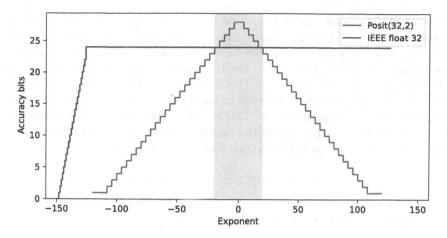

Fig. 2. Accuracy binary digits for 32-bit formats

$$X = (-1)^s \times (useed)^k \times 2^e \times (1 + f), \tag{1}$$

where $useed = 2^{2^{es}}$, e is the integer encoded by the exponent field, k is $l-1$ when $r = 1$, or $-l$ when $r = 0$, and f is the normalized fraction (this is, the value encoded by the fraction bits divided by 2^F, so $0 \le f < 1$). Under this decoding approach, if a value is negative (when the sign bit is 1), its two's complement is computed before extracting the regime, exponent, and fraction, so values k, e and f in Eq. (1) are always considered from the absolute value of the posit.

The main differences with the standard floating-point format are the utilization of an unsigned and unbiased exponent, the hidden bit of the significand is always "1" (no subnormal numbers are considered), and the existence of the variable-length regime field. However, notice that this decoding is quite similar to the one for classical floating-point numbers: it deals with a sign bit, a signed exponent (regime and exponent can be gathered in a single factor) and a significand with a hidden bit. As a consequence, the circuit design for both arithmetic formats would be similar too. In fact, this float-like decoding scheme is the one used by most of the posit arithmetic units from the literature [4,11,16], as well as by the approximate posit units proposed so far [18,20].

Apparently, trying to implement posits by first forcing them to look more like floats and then converting back does not seem optimal, and the community is still in the early stages of discovering new decodings and circuit shortcuts that leverage this recently proposed format.

2.2 Two's Complement Posit Decoding

The previous decoding scheme of posit numbers deals with negative numbers in a similar manner as signed integers do. From a hardware perspective, converting posits to their absolute value before decoding them adds extra area and performance overhead, specially when compared with IEEE 754 floats. To address this issue, Isaac Yonemoto, co-author of [8], proposes a different way of decoding posit numbers: for negative values, the most significant digit of the significand is treated as "−2" instead of "1". The rest of the fields remain the same, but under this approach there is no need to compute the two's complement (absolute value) of each negative posit. This is consistent with the way posits were initially intended, as a mapping of the signed (two's complement) integers to the projective reals. The value X of a posit number is now given by (2)

$$X = (useed)^{\tilde{k}} \times 2^{\tilde{e}} \times (1 - 3s + f), \tag{2}$$

where again $useed = 2^{2^{es}}$, but now $\tilde{e}$ is equal to e XOR-ed bitwise with s and $\tilde{k}$ is $-l$ when $r = s$, or $l - 1$ otherwise.

Theorem 1. *For any given posit bit string that encodes a number other than zero or NaR, the expressions* (1) *and* (2) *are equivalent.*

Proof. When the sign bit is 0 (i.e. the bit string encodes a positive number), it is trivial that both expressions evaluate the same.

On the other hand, the case when $s = 1$ requires more attention. First, note that in such a case, two's complement of the bit string must be computed before evaluating expression (1). Hence, all the bits at the left of the rightmost "1" are inverted. Let us consider three cases, depending on which field that bit belongs to.

(i) If the rightmost "1" bit belongs to the fraction field, the fraction $f \neq 0$. Hence, it is evident that $k = \tilde{k}$ and $e = \tilde{e}$, since k and e are obtained from the inverted regime and exponent bits in the original bit string, respectively. It remains to check whether the significands from expressions (1) and (2) have an opposite value. But recall that the fraction from expression (1) is two's complemented, and due to the fact that $0 \leq f < 1$, the two's complement of f is $\tilde{f} = 1 - f$. From this last property it follows that $(1 + \tilde{f}) = -(-2 + f)$.

(ii) If the rightmost "1" bit belongs to the exponent field, then $f = 0$ and $e \neq 0$. For the same reason as in the previous case, $k = \tilde{k}$. But now the exponent field for expression (1) is two's complemented rather than inverted, so we have that $e = \tilde{e} + 1$. However, since $f = 0$, the significand in expression (1) evaluates 1, while expression (2) evaluates $(1 - 3s + f)$ as -2, which compensates the difference in the exponents.

(iii) If the rightmost "1" bit belongs to the regime, then $e = f = 0$. Also, it should be noted that such a bit corresponds to the last (inverted) regime bit (in case the regime is a sequence of 0's) or to the bit immediately preceding the inverted one (when the regime is a sequence of 1's). In both cases, taking two's complement for computing (1) reduces the length of the regime field in 1, so it follows that $k = \tilde{k} + 1$. In addition, note that in this situation $e = 0$, while $\tilde{e} = 2^{es} - 1$. Nevertheless, a similar situation as in the previous case occurs with the fraction: the significand from expression (2) is evaluated as -2, which compensates the difference of exponents previously mentioned. Note that the multiplicands of both expressions are powers of 2, so it suffices to check that both expressions have the same exponent. Indeed: $(2^{es})^{\tilde{k}} + \tilde{e} + 1 = (2^{es})^{k-1} + (2^{es} - 1) + 1 = (2^{es})^k$. □

When dealing with the hardware implementation, the significand of (1) can be represented in fixed-point with a single (hidden) bit that always takes the value "1". On the other hand, when considering expression (2), the significand $(1 - 3s + f)$ belongs to the interval $[-2, -1)$ for negative posits and to $[1, 2)$ for positive ones, so such signed fixed-point representation requires two integer (or hidden) bits that depend on the sign of the posit. More precisely, in this case, negative posits prepend "10" to the fraction bits as the 2's complement hidden bits, and positive posits prepend "01". Note how this contrasts with the unsigned fixed-point representation of the significand in the floating-point and classical sign-magnitude posit decoding formats. Therefore, this approach eliminates complexity in the decoding and encoding stages, but requires redesigning some of the logic when implementing posit operators.

There are not many works that implement this two's complement decoding approach for posit numbers. The first implementation of posit adders and multipliers based on this decoding appeared in [24], and more details about such a scheme were introduced in [7]. Also, [19] presents different energy-efficient fused posit MAC units that follow the same approach as [24].

In this paper we present a generic implementation of posit functional units based on two's complement decoding. Furthermore, we compare different state-of-the-art posit units based on both decoding schemes.

Finally, it is noteworthy that previous works have examined the effect of two's complement notations in floating-point arithmetic [1]. However, in such a case, some features or properties are lost with respect to the IEEE standard for floats. In this work we prove that both sign-magnitude and two's complement coding of posit numbers are equivalent, and therefore all properties are preserved regardless of the used approach. The impact of each decoding scheme is found on the hardware implementation, as will be discussed in Sect. 4.

3   Posit Operators

The advantage of using sign-magnitude decoding for posit numbers is that arithmetic operations can be performed in a similar way to standard floating-point ones (except for the bitwidth of the fields and exception handling). While this can leverage the already designed circuits for floating point, forcing posits to look like floats and then converting back adds some overhead to the operators. However, considering the posit significand as a signed fixed-point value eliminates the need for absolute value conversion, but requires some redesign of the arithmetic cores.

This section describes in detail and compares the design of different arithmetic operations when dealing with each posit decoding scheme.

3.1 Decoding and Encoding Stages

Unlike floating-point hardware that ignores subnormal numbers, the variable-length regime does not allow the parallel decoding of posit numbers, that is, the fraction and exponent cannot be extracted until the length of the regime is known. Thus, when implementing posit operators in hardware, it is usually necessary to extract the four fields presented in Sect. 2 (s, k, e and f, plus a flag for zero/NaR exceptions) from a compact posit number before starting the real computation, as well as packing again the resulting fields after that. The components that perform such processes are usually known as decoders and encoders, respectively, and those are the modules that present more differences in their design according to the decoding mode used.

The classical decoding scheme considers negative posits to be in two's complement. Hence, in such a case, it is necessary to first take a two's complement of the remaining bit string before decoding the regime (which is usually done with a leading ones/zeros detector), exponent and fraction bits, as detailed in Algorithm 1 (zero/NaR exception checking is omitted for the sake of clarity). Then, all the computation is performed with the absolute value of the posits, leaving aside the sign logic until the end, where it requires to take again the two's complement of the bit string according to the sign of the result. The process of encoding a posit from its different fields mainly consists of performing Algorithm 1 backwards, plus handling possible rounding and overflow/underflow situations.

Algorithm 1 Classical posit decoding algorithm

Require: $X \in \text{Posit}\langle n, es \rangle$, $F = n - es - 3$
Ensure: $(-1)^s \times (useed)^k \times 2^e \times (1 + f) = X$
  $s \leftarrow X[n-1]$
  if $s = 1$ **then**
    $p \leftarrow \sim X[n-2:0] + 1$ ▷ Take two's complement
  else if $s = 0$ **then**
    $p \leftarrow X[n-2:0]$
  end if
  $r \leftarrow p[n-2]$
  $l \leftarrow LZOC(p)$ ▷ Count regime length
  if $r = 1$ **then**
    $k \leftarrow l - 1$
  else if $r = 0$ **then**
    $k \leftarrow -l$
  end if
  $q \leftarrow p[n-l-3:0]$ ▷ Extend with 0's to the right, if necessary
  $e \leftarrow q[F + es - 1 : F]$
  $f \leftarrow q[F - 1 : 0]$

On the other hand, the alternative scheme proposed by Yonemoto handles both positive and negative posit numbers simultaneously, without the need of computing the absolute value of the posits. Determining the sign of the regime's value requires checking if the posit sign bit is equal to the MSB of the regime, and the exponent value for this case requires XOR-ing es bits with the sign bit. The decoding process for this approach is described in Algorithm 2. Also, as already mentioned, handling the significand in signed fixed-point format for computation requires one extra bit for the sign, since this decoding considers the significand of positive values to be in the interval $[1, 2)$ (as in the previous decoding), or in $[-2, -1)$ when the number is negative. This approach can reduce the latency of the decoding and encoding stages, specially for larger bitwidths, since it requires XOR-ing just es bits (generally es is not greater than 3) instead of computing the two's complement of the n-bit posits as in Algorithm 1. However, the signed fixed-point significand introduces extra complexity in the core of the arithmetic operations, as will be discussed below.

Algorithm 2 Alternative posit decoding algorithm

Require: $X \in \text{Posit}\langle n, es \rangle$, $F = n - es - 3$
Ensure: $(useed)^k \times 2^e \times (1 - 3s + f) = X$

    $s \leftarrow X[n-1]$
    $p \leftarrow X[n-2:0]$
    $r \leftarrow p[n-2]$
    $l \leftarrow LZOC(p)$ ▷ Count regime length
    if $r \neq s$ **then**
        $k \leftarrow l - 1$
    else if $r = s$ **then**
        $k \leftarrow -l$
    end if
    $q \leftarrow p[n - l - 3 : 0]$ ▷ Extend with 0's to the right, if necessary
    $e \leftarrow q[F + es - 1 : F] \oplus \{es\{s\}\}$ ▷ Sign bit is replicated to perform XOR
    $f \leftarrow q[F - 1 : 0]$

3.2 Addition

In posit arithmetic, as well as in the case of floating-point arithmetic, when performing the addition (or subtraction) of two numbers, it is necessary to shift one of the fractions so both exponents are equal. If the first exponent is smaller than the second, the first fraction is shifted to the right by a number of bits given by the absolute difference of the exponents. Otherwise, the same is done to the second fraction. Then, the aligned significands are added, and the result is normalized, so the larger exponent is adjusted if needed.

When using a classical sign-magnitude decoding approach, some extra logic is needed to handle the sign of the result. But such logic is eliminated when dealing with signed significands, since the sign of the result can be inferred from the leftmost bit of the addition of the significands, without initially comparing the magnitude of the inputs. Dealing with signed significands also avoids the need of performing subtraction or taking two's complement when the sign of both addends differ, which makes up for using one extra bit in the addition of significands. On the other hand, normalization of the significand in two's complement deserves special attention, since it needs to count not only the leading zeroes, but the leading ones when the result is negative. However, as will be shown in Sect. 4, this does not involve hardware overhead for this particular module.

3.3 Multiplication

In a similar manner as in the case of addition, posit multiplication takes inspiration from the floating-point algorithm: both significands are multiplied and normalized, and the exponents are added together. Additionally, the result from significand multiplication must be normalized to fit in the corresponding interval, which involves shifting the fraction plus adding to the exponent the number of shifted bits.

When the significands have a single hidden bit, i.e., using the sign-magnitude posit decoding, the leftmost bit of the multiplication indicates if the resulting fraction must be shifted and 1 must be added to the exponent. Note that in this case, both multiplicands follow the expression $(1 + f) \in [1, 2)$, so the product must be in the interval $[1, 4)$. Thus, normalizing the result might require shifting one bit at most.

On the other hand, when dealing with the two's complement decoding scheme, even though the multiplication can be performed directly (just one more bit for each operand is necessary), the normalization of the result is more complex in this case. As each multiplicand $(1 - 3s + f)$ can be in the range $[-2, -1) \cup [1, 2)$, the result will fall within the range $(-4, -1) \cup [1, 4]$. In terms of fixed-point arithmetic, the operand has two integer bits, so the multiplication has four bits that represent the integer part of the number and that should be examined in the normalization process. Note that the resulting sign is also implicit in the multiplication result. However, this approach introduces one extra case that needs special attention: when the two multiplicands are equal to -2, the result is 4, which requires adding 2, rather than 1, to the exponent when normalizing the result.

Finally, note that similar considerations should be taken into account for the case of the division operation, although it is beyond the scope of this paper.

4 Hardware Evaluation

This section evaluates the hardware impact of each posit decoding scheme. In addition to standard comparison of arithmetic units, in order to provide a more fine-grained evaluation, this section compares the hardware requirements of each individual component when using each of the decodings. To achieve an accurate evaluation, all the results given in this work were generated to be purely combinational and synthesized targeting a 45 nm TSMC standard-cell library with no timing constraint and typical case parameters using Synopsys Design Compiler.

4.1 Components Evaluation

To have a better understanding of how the different ways of decoding posit numbers impact on the hardware resource utilization, we compared ASIC synthesis results for each single component of the posit operators described in Sect. 3 when implemented under each of the decoding schemes presented in this paper. For the sign-magnitude decoding, we extracted the different components from Flo-Posit [16], which includes open-source[1] designs implemented using FloPoCo [5] and requires less hardware resources than other implementations based on the same classical decoding scheme. With respect to the two's complement decoding, there are no available designs other than those proposed in [24], which consist of a C++ header library for HLS that implement a modified posit format. Thus, in order to make as fair a comparison as possible, we implemented the designs described in Sect. 3 using Yonemoto's decoding scheme and using FloPoCo as well, which allows to generate parameterized units for any number of bits and exponent size.

In order to verify the correctness of the proposed architectures, exhaustive tests for units with 16 bits or less, as well as random tests with corner cases for larger bitwidths, were performed using a VHDL simulator. The results were compared against two software libraries: the Universal number library [21], which supports arithmetic operations for any arbitrary posit configuration, and GNU MPFR, which was modified with support for posit binary representation. All these tests were successful. Then, each module (decoder, encoder, core adder and core multiplier) was synthesized separately, so the area, power, datapath delay and energy (power-delay product) could be compared in detail. Results were normalized with respect to the classical decoding scheme.

As can be seen in Fig. 3, which shows the cost of just the decoding stage rather than the cost of the whole arithmetic operation, considerable savings are obtained when decoding the posits by using Yonemoto's two's complement proposal. Under this approach, the decoder module requires about 66% of the area, 45% of power, 59% of delay and 27% of the energy than the same module implemented using the classical sign-magnitude decoding approach. Also, it is noteworthy that for many of the most common operations, like addition or multiplication, two operands need to be decoded, so this module is often duplicated.

Similar figures are obtained for the encoder module. As Fig. 4 shows, using Yonemoto's approach requires about 33% less area than the classical one, but in this case the power and delay savings are not as pronounced as for the decoder module. Nevertheless, using the alternative decoding scheme reduces energy consumption of this process by half.

As already mentioned, dealing with signed significands avoids the need of negating one of the operands when performing addition of different sign values. This is demonstrated in Fig. 5, which compares the hardware requirements of both approaches for just the logic of posit addition (without circuitry for decoding operands nor for rounding/encoding the result). The extra bit for dealing

---

[1] https://github.com/artecs-group/Flo-Posit/tree/6fd1776.

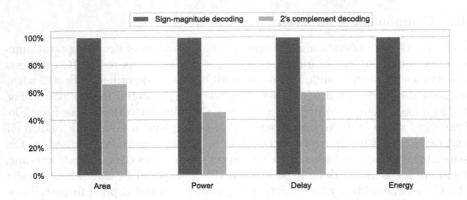

Fig. 3. Relative hardware performance metrics of Posit$\langle 32, 2 \rangle$ decoder components.

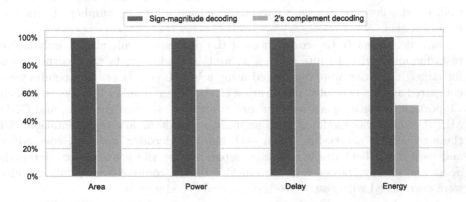

Fig. 4. Relative hardware performance metrics of Posit$\langle 32, 2 \rangle$ encoder components.

with the significand in two's complement adds negligible overhead to this component, and together with the reduction of logic to handle addition of different sign operands, makes this scheme to use 91% of the area, 86% of the power and 84% of the datapath delay of the analogous component based on the classical decoding.

The case of the multiplier module is different from the previous ones. Here, handling the significands in two's complement requires one extra bit for each operand, and a total of four more bits for storing the multiplication result, when compared with the sign-magnitude approach. This translates into approximately 7% more area and power, but similar delay, as shown in Fig. 6.

4.2 Comparison with the State-of-the-Art

The components using two's complement decoding scheme seem to provide smaller and faster implementations. However, it is important to verify that whole operators follow the same trend, and that the proposed implementations are not

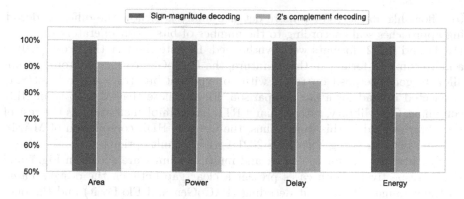

Fig. 5. Relative hardware performance metrics of Posit$\langle 32, 2 \rangle$ adder components

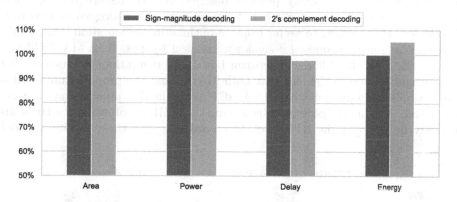

Fig. 6. Relative hardware performance metrics of Posit$\langle 32, 2 \rangle$ multiplier components.

sub-optimal. For this purpose, three different implementations of posit operators from the state-of-the-art were compared: PACoGen [11][2] and Flo-Posit [16], which use the sign-magnitude decoding scheme given by (1), and MArTo [24][3], which is based in the decoding scheme proposed by Yonemoto with slight differences. In particular, the designs presented in [24] perform conversion to/from the so-called *posit intermediate format* (PIF), a custom floating-point format that stores the significand in two's complement (just like the approach evaluated in this work) and takes an exponent (including the regime) which is biased with respect to the minimum exponent, as in the IEEE 754 standard. The PIF simplifies the critical path of the operators, at the cost of small additions in the decoding/encoding of posits. Also, the proposed implementation that has been discussed in the preceding section was added to the comparison, so there are two implementations for each posit decoding approach. Unlike MArTo, the proposed operators implement the logic in pure posit format, without conversion of posits

---

[2] https://github.com/manish-kj/PACoGen/tree/5f6572c.
[3] https://gitlab.inria.fr/lforget/marto/tree/2f053a56.

to a float-like format. For a more detailed analysis of how the different decoding approaches scale according to the number of bits, posit operators for $\langle 8, 1 \rangle$, $\langle 16, 1 \rangle$ and $\langle 32, 2 \rangle$ formats were synthesized. Despite the fact that Posit$\langle 8, 0 \rangle$ is a more common format in the literature, the PACoGen core generator does not allow to generate posit operators with no exponent bits ($es = 0$), so Posit$\langle 8, 1 \rangle$ is selected instead for a fair comparison. Finally, note that MArTo is an HLS-compliant C++ library, rather than a RTL-based implementation as the rest of the libraries used in this work. Thus, the C++ to HDL compilation of MArTo operators is done using Vitis HLS 2021.1 with default options.

Synthesis results for the adder and multiplier units are shown in Fig. 7 and Fig. 8, respectively. Both cases present a clear gap between the designs based on the sign-magnitude posit decoding (PACoGen and Flo-Posit) and the ones using the two's complement scheme (MArTo and the one proposed in this paper), specially for the power-delay product (energy) results. Except for the adder delay, Flo-Posit designs present better figures than the analogous designs from PACoGen, which seems to be far from an optimal implementation. Exactly the same occurs for the proposed designs with respect to those from MArTo library, but in this case the difference between both implementations is much smaller. This might be due to the fact that both MArTo and the proposed units follow quite similar designs but with certain differences in the implementation, since the former designs are generated by a commercial HLS tool, while the latter are directly designed at the RTL level. In accordance with these results, we took

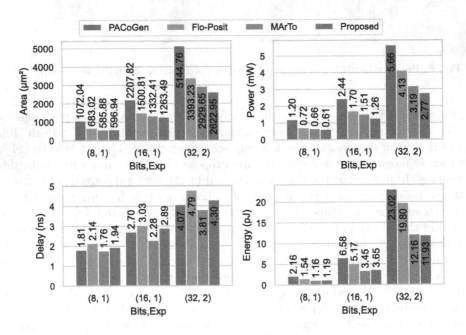

Fig. 7. Synthesis results for different Posit$\langle n, es \rangle$ adder designs.

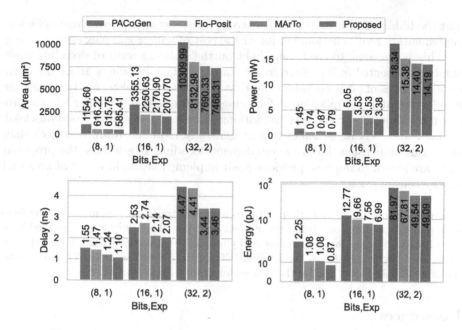

Fig. 8. Synthesis results for different Posit$\langle n, es \rangle$ multiplier designs.

the best designs of each decoding scheme (Flo-Posit and the proposed one) as a baseline for detailed comparison.

In the case of posit adders, the greatest resource savings are obtained for 32-bit operators: when using the alternative decoding, the area is reduced by 22.70%, the power by 32.90%, the delay by 10.22% and the energy by 39.77%.

On the other hand, and in line with the results shown previously, the alternative two's complement decoding scheme for posit multiplier units also presents less resource utilization when compared with the classical float-like scheme, but these savings are not as pronounced as in the case of posit addition. As can be seen in Fig. 8, the 32-bit multipliers based on Yonemoto's decoding approach reduce area, power, datapath delay and energy by 8.17%, 7.72%, 21.54%, and 27.60%, respectively.

5   Conclusions

Multiple designs of posit arithmetic units have been proposed since the appearance of this alternative format. While those units might present several optimizations for area or energy efficiency, one of the main design differences is the way posit strings are decoded. The first works on posit arithmetic presented a sign-magnitude decoding scheme similar to floating-point arithmetic, with a sign bit, a signed exponent and a fraction with a hidden bit equal to 1. But recently, a two's complement decoding for posits proposed by I. Yonemoto, which considers that the hidden bit means -2 for negative posits, seems to provide more efficient implementations of functional units, at the cost of a more complex circuit design.

This paper aims to shed some light on the different ways of decoding posit numbers presented in literature so far, and how such decodings affect the hardware resources of posit operators. To that purpose, we implemented custom-size posit adder and multiplier units following the alternative decoding scheme proposed by Yonemoto. Synthesis evaluations show that posit units based on classical float-like decoding schemes require generally more hardware resources than analogous units using the two's complement decoding. In addition, the proposed units are shown to improve previous posit implementations in terms of area and energy consumption.

Acknowledgments. The authors wish to acknowledge Isaac Yonemoto for the feedback and explanations on his insight of posit decoding. This work was supported by a 2020 Leonardo Grant for Researchers and Cultural Creators, from BBVA Foundation, whose id is PR2003_20/01, by the EU(FEDER) and the Spanish MINECO under grant RTI2018-093684-B-I00, and by the CM under grant S2018/TCS-4423.

References

1. Boldo, S., Daumas, M.: Properties of two's complement floating point notations. Int. J. Softw. Tools Technol. Transf., 237–246 (2003). https://doi.org/10.1007/s10009-003-0120-y
2. Camus, V., Enz, C., Verhelst, M.: Survey of precision-scalable multiply-accumulate units for neural-network processing. In: 2019 IEEE International Conference on Artificial Intelligence Circuits and Systems (AICAS), pp. 57–61. IEEE, March 2019. https://doi.org/10.1109/AICAS.2019.8771610
3. Carmichael, Z., Langroudi, H.F., Khazanov, C., Lillie, J., Gustafson, J.L., Kudithipudi, D.: Deep positron: a deep neural network using the posit number system. In: 2019 Design, Automation & Test in Europe Conference & Exhibition (DATE), pp. 1421–1426. IEEE, March 2019. https://doi.org/10.23919/DATE.2019.8715262
4. Chaurasiya, R., et al.: Parameterized posit arithmetic hardware generator. In: 2018 IEEE 36th International Conference on Computer Design (ICCD), pp. 334–341. IEEE (2018). https://doi.org/10.1109/ICCD.2018.00057
5. de Dinechin, F., Pasca, B.: Designing custom arithmetic data paths with FloPoCo. IEEE Des. Test Comput. **28**(4), 18–27 (2011). https://doi.org/10.1109/MDT.2011.44
6. Goldberg, D.: What every computer scientist should know about floating-point arithmetic. ACM Comput. Surv. (CSUR) **23**(1), 5–48 (1991). https://doi.org/10.1145/103162.103163
7. Guntoro, A., et al.: Next generation arithmetic for edge computing. In: 2020 Design, Automation & Test in Europe Conference & Exhibition (DATE), pp. 1357–1365. IEEE (2020). https://doi.org/10.23919/DATE48585.2020.9116196
8. Gustafson, J.L., Yonemoto, I.T.: Beating floating point at its own game: posit arithmetic. Supercomput. Frontiers Innov. **4**(2), 71–86 (2017). https://doi.org/10.14529/jsfi170206

9. Ho, N.m., Nguyen, D.T., Silva, H.D., Gustafson, J.L., Wong, W.F., Chang, I.J.: Posit arithmetic for the training and deployment of generative adversarial networks. In: 2021 Design, Automation & Test in Europe Conference & Exhibition (DATE), pp. 1350–1355. IEEE, February 2021. https://doi.org/10.23919/DATE51398.2021.9473933

10. IEEE Computer Society: IEEE Standard for Floating-Point Arithmetic. IEEE Std 754-2019 (Revision of IEEE 754-2008), pp. 1–84 (2019). https://doi.org/10.1109/IEEESTD.2019.8766229

11. Jaiswal, M.K., So, H.K.: PACoGen: a hardware posit arithmetic core generator. IEEE Access **7**, 74586–74601 (2019). https://doi.org/10.1109/ACCESS.2019.2920936

12. Johnson, J.: Rethinking floating point for deep learning. arXiv e-prints, November 2018

13. Lindstrom, P.: Variable-radix coding of the reals. In: 2020 IEEE 27th Symposium on Computer Arithmetic (ARITH), pp. 111–116. IEEE, June 2020. https://doi.org/10.1109/ARITH48897.2020.00024

14. Mallasén, D., Murillo, R., Del Barrio, A.A., Botella, G., Piñuel, L., Prieto, M.: PERCIVAL: open-source posit RISC-V core with quire capability, pp. 1–11, November 2021

15. Muller, J.M., et al.: Handbook of Floating-Point Arithmetic. Birkhäuser Boston, Boston (2010). https://doi.org/10.1007/978-0-8176-4705-6

16. Murillo, R., Del Barrio, A.A., Botella, G.: Customized posit adders and multipliers using the FloPoCo core generator. In: 2020 IEEE International Symposium on Circuits and Systems (ISCAS), pp. 1–5. IEEE, October 2020. https://doi.org/10.1109/iscas45731.2020.9180771

17. Murillo, R., Del Barrio, A.A., Botella, G.: Deep PeNSieve: a deep learning framework based on the posit number system. Digit. Sig. Process. Rev. J. **102**, 102762 (2020). https://doi.org/10.1016/j.dsp.2020.102762

18. Murillo, R., Del Barrio Garcia, A.A., Botella, G., Kim, M.S., Kim, H., Bagherzadeh, N.: PLAM: a posit logarithm-approximate multiplier. IEEE Trans. Emerging Top. Comput., 1–7 (2021). https://doi.org/10.1109/TETC.2021.3109127

19. Murillo, R., Mallasén, D., Del Barrio, A.A., Botella, G.: Energy-efficient MAC units for fused posit arithmetic. In: 2021 IEEE 39th International Conference on Computer Design (ICCD), pp. 138–145. IEEE, October 2021. https://doi.org/10.1109/ICCD53106.2021.00032

20. Norris, C.J., Kim, S.: An approximate and iterative posit multiplier architecture for FPGAs. In: 2021 IEEE International Symposium on Circuits and Systems (ISCAS), pp. 1–5. IEEE, May 2021. https://doi.org/10.1109/ISCAS51556.2021.9401158

21. Omtzigt, T., Gottschling, P., Seligman, M., Zorn, B.: Universal numbers library: design and implementation of a high-performance reproducible number systems library. arXiv e-prints (2020)

22. Sharma, N., et al.: CLARINET: a RISC-V based framework for posit arithmetic empiricism, pp. 1–18, May 2020

23. Sze, V., Chen, Y.H., Yang, T.J., Emer, J.S.: Efficient processing of deep neural networks: a tutorial and survey. Proc. IEEE **105**(12), 2295–2329 (2017). https://doi.org/10.1109/JPROC.2017.2761740

24. Uguen, Y., Forget, L., de Dinechin, F.: Evaluating the hardware cost of the posit number system. In: 2019 29th International Conference on Field Programmable Logic and Applications (FPL), pp. 106–113. IEEE (2019). https://doi.org/10.1109/FPL.2019.00026

Universal: Reliable, Reproducible, and Energy-Efficient Numerics

E. Theodore L. Omtzigt[1]([✉]) [iD] and James Quinlan[2] [iD]

[1] Stillwater Supercomputing, Inc., El Dorado Hills, CA 95762, USA
tomtzigt@stillwater-sc.com
[2] University of New England, Biddeford, ME 04005, USA
jquinlan@une.edu
https://stillwater-sc.com

Abstract. *Universal* provides a collection of arithmetic types, tools, and techniques for performant, reliable, reproducible, and energy-efficient algorithm design and optimization. The library contains a full spectrum of custom arithmetic data types ranging from memory-efficient fixed-size arbitrary precision integers, fixed-points, regular and tapered floating-points, logarithmic, faithful, and interval arithmetic, to adaptive precision integer, decimal, rational, and floating-point arithmetic. All arithmetic types share a common control interface to set and query bits to simplify numerical verification algorithms. The library can be used to create mixed-precision algorithms that minimize the energy consumption of essential algorithms in embedded intelligence and high-performance computing. *Universal* contains command-line tools to help visualize and interrogate the encoding and decoding of numeric values in all the available types. Finally, *Universal* provides error-free transforms for floating-point and reproducible computation and linear algebra through user-defined rounding techniques.

Keywords: Mixed-precision algorithm · Energy-efficient arithmetic · Reliable computing · Reproducibility

1 Introduction

The proliferation of computing across more diverse use cases drives application and algorithm innovation in more varied goals. Performance continues to be a key driver in product differentiation, but energy efficiency is essential for embedded and edge computing. Moreover, reliable and reproducible computing is paramount in safety applications such as autonomous vehicles.

In embedded systems, energy is at a premium, and the application must deliver a solution within a strict power constraint to be viable. In hyperscaled

---

Developed by open-source developers, and supported and maintained by Stillwater Supercomputing Inc.

J. Gustafson and V. Dimitrov (Eds.): CoNGA 2022, LNCS 13253, pp. 100–116, 2022.
https://doi.org/10.1007/978-3-031-09779-9_7

cloud data centers, the cost of electricity has overtaken the acquisition cost of IT equipment, making energy efficiency even an economic driver for the cloud.

The reproducibility of computational results is essential to applications that impact human safety and collaboration. Reproducible computation is required for forensic analysis to explain a recorded failure of an autonomous vehicle. Collaborative projects that leverage computational science are more efficient when two different research groups can reproduce simulation results on different platforms. *Universal* provides the deferred rounding machinery to implement reproducibility. Lastly, numerically sensitive results require verification or quality assertions. Reliable computing provides such guarantees on accuracy or bounding boxes of error.

The *Universal* library offers custom arithmetic types and utilities for optimizing energy efficiency, performance, reproducibility, and reliability of computational systems. Once the arithmetic solution has been found, *Universal* provides a seamless transition to create and leverage custom compute engines to accelerate the execution of the custom arithmetic. And finally, *Universal* provides a unified mechanism to extend other language environments, such as MATLAB/Simulink, Python, or Julia, with validated and verified custom arithmetic types.

2   Background and Motivation

Deep learning algorithms transform applications that classify and characterize patterns in vision, speech, language, and optimal control. This so-called Software 2.0 transformation uses data and computation to synthesize and manage the behavior and capability of the application. In deep learning applications, computational demand is high, and data supplied is varied, making performance and energy efficiency paramount for success. The industry has responded with a proliferation of custom hardware accelerators running energy-conserving numeric systems.

These hardware accelerators need to be integrated into embedded, network, and cloud ecosystems. For example, Google TPUs [16], and Intel CPUs [15] support a type called a brain float, which is a 16-bit floating-point format that truncates the lower 16-bits of a standard IEEE-754 single-precision float. NVIDIA, on the other hand, implements their unique data type, the TensorFloat-32 (TF32) [17], which is a 19-bit format with 8 bits encoding the exponent, and 10 bits encoding the mantissa.

This proliferation of vendor-specific types creates demand for solutions that enable software designers to create, run, test, and deploy applications that take advantage of custom arithmetic while, at the same time, integrating seamlessly with a broad range of hardware accelerators. Software that can evaluate, adapt, or replace different arithmetic types requires an upfront investment in architecture design and implementation. *Universal* offers an environment where anyone can deploy custom arithmetic types while maintaining complete flexibility to adapt to better hardware from a different vendor when available.

As the research and development community learns more about the computational dynamics of Software 2.0 applications, novel number system representations will be invented to enable new application capabilities. For example, massive multiple-input multiple-output (MIMO) systems in cellular networks will benefit from optimized arithmetic [21]. Convolutional Neural Networks are showing attributes that favor logarithmic number systems [27]. Safety systems require arithmetic that is reproducible and numeric algorithms that are reliable [22]. Large scale high-performance computing applications that model physical phenomena leverage continuity constraints to compress fields of metrics to lower power consumption and maximize memory performance [13,18] This list of custom number systems and their arithmetic type representations will only grow, igniting a renewed focus on efficiency of representation and computation.

Numerical environments such as `Boost MultiPrecision` [19], `MPFR` [5], and `GMP` [7] have been focused on providing extensions to IEEE-754. They are not tailored to providing new arithmetic types and encodings as required for improving Software 2.0 applications. In contrast, *Universal* is purposefully designed to offer and integrate new arithmetic types for the emerging applications in embedded intelligence, mobile, and cloud computing.

Universal started in 2017 as a hardware verification library for the emerging posit standard [9]. It provided a hardware model of a bit-level implementation of arbitrary configuration posits, parameterized as `posit<nbits,es>` and presented as a plug-in arithmetic type for C++ linear-algebra libraries [6,25,26]. Since then, *Universal* has grown into a research and development platform for multi-precision algorithm optimization and numerical analysis.

More recently, it has been instrumental in developing applications that exhibit strong cooperation between general-purpose processing on the CPU and special-purpose processing on accelerators. As *Universal* arithmetic types operate with the same encoding and memory layout as the hardware accelerator, applications can use the general-purpose CPU to serialize, prepare, and manage the data structures on behalf of the custom hardware accelerator without the need for conversions to and from native types.

In this paper, we provide a status update to the third edition of the *Universal* library [23]. In Sect. 3, we discuss the various arithmetic types available in *Universal* and where they fit in the set hierarchy representing abstract algebraic number systems. Section 4 discusses the design flow for creating new arithmetic types that have proven to be productive. *Universal* is still very much a hardware/software co-design library, so Sect. 5 describes the standard application programming interface of the arithmetic types in *Universal* that simplify integration into verification and regression test suites. Finally, Sect. 6 demonstrates different algorithm and application framework integration examples. We conclude in Sect. 7 with a summary and future work.

3   Universal Arithmetic Type Organization

Figure 1 shows the cover of the different arithmetic types available in *Universal* relative to the known algebraic sets.

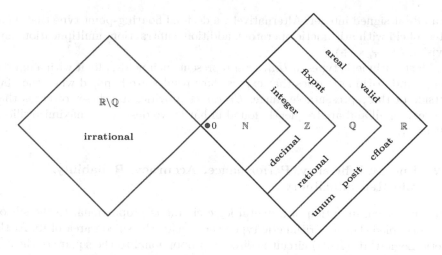

Fig. 1. Abstract algebraic sets and *Universal* arithmetic types

Universal classifies arithmetic types into *fixed* and *adaptive* types. *Fixed* types are arithmetic types that have a fixed memory layout when declared. In contrast, the memory layout of *adaptive* types varies during the computation. Fixed types are intended for energy-efficient, performant, and linear-algebra-focused applications. Adaptive types are more suitable for accuracy and reliable computing investigations.

For the *fixed* arithmetic types, *Universal* strives to offer sizes that are configurable by individual bits as the target are custom hardware implementations in specialized hardware accelerators. The parameterization space of *fixed* arithmetic types is:

1. sampling profile and encoding
2. size in bits
3. dynamic range
4. arithmetic behavior (modulo, saturate, clip etc.)

The most informative example is the classic floating-point type: **cfloat**. It is parameterized in all dimensions:

```
cfloat<nbits, es, BlockType,
        hasSubnormals, hasSupernormals, isSaturating>
```

We will discuss the details in Sect. 3.3.

3.1 Definitions

In *Universal* a custom arithmetic type is defined by a memory layout of the data type, an encoding, and a set of operators that approximate an abstract algebra. For example, a simple algebra on a ring where the data encoding is constrained to an 8-bit signed integer. Alternatively, a decimal floating-point type that emulates a field with arithmetic operators, addition, subtraction, multiplication, and division $(+, -, \times, \div)$.

Most arithmetic types in *Universal* represent such fields, albeit with a limited range, and arithmetic rules that express how results are handled when they fall outside of the representable range. *Universal* provides a rich set of types that are very small and are frequently found in hardware designs to maximize silicon efficiency.

3.2 Energy Efficiency, Performance, Accuracy, Reliability, and Reproducibility

The energy consumption of a digital logic circuit is proportional to the silicon area occupied. For an arithmetic type over a field, the silicon area of its Arithmetic Logic Unit (ALU) circuit is directly proportional to the square of the size of the encoding. Therefore, the constraint on the range covered by an arithmetic type is a crucial design parameter for energy efficiency optimization.

However, in sub-micron chip manufacturing technology, the energy consumption of data movement is more significant [14]. In 45 nm technology, an 8-bit addition consumes about 0.03pJ, and a 32-bit floating-point multiply consumes 0.9pJ. However, reading a 32-bit operand from a register file is 5pJ, and reading that same word from external memory is 640pJ. This discrepancy of energy consumption between operator and data movement worsens with smaller manufacturing geometries, adding additional importance to maximize information content in arithmetic types to make them as dense and small as possible.

The performance of arithmetic types is also strongly impacted by data movement. Any memory-bound algorithm will benefit from moving fewer bits to and from external memory. Therefore, arithmetic types and algorithms must be co-designed to minimize the number of bits per operation to maximize performance.

For applications constrained by accuracy, such as simulation and optimization, arithmetic types need to cover the precision and dynamic range of the computation. The precision of a type is the difference between successive values representable by its encoding. The dynamic range is the difference between the smallest and the largest value representable by the encoding.

The domain of reliable numerical computing must provide verified answers. Arithmetic types for reliable computing need to guarantee numerical properties of the result [4]. For example, interval arithmetic can assert that the true computation answer lies in some interval and is an example of reliable computing.

Reproducible computation guarantees that results are the same regardless of the execution order. Reproducible computing is particularly pertinent for high concurrency environments, such as GPUs and High-Performance Computing (HPC) clusters.

3.3 Fixed Size, Arbitrary Precision

Fixed-size, arbitrary precision arithmetic types are tailored to energy-efficiency
and memory bandwidth optimization.

integer<nbits, BlockType, NumberType> The **integer** arithmetic type can
  represent Natural Numbers, Whole Numbers, and integers of *nbits*. Natural and Whole Numbers are encoded as 1's complement numbers, and the
  Integers are encoded as a 2's complement numbers. Figure 2 shows a 16-
  bit incarnation. Its closure semantics are modulo, and effectively extend the
  C++ language with arbitrary precision signed integers of arbitrary fixed-size.
  This type is very common in hardware designs.

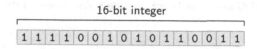

Fig. 2. A 16-bit integer.

fixpnt<nbits, rbits, arithmetic, BlockType> The **fixpnt** arithmetic
  type is a 2's complement encoded fixed-point of *nbits* with the radix point set
  at bit *rbits*. Figure 3 shows a 16-bit incarnation with the radix point at bit
  8. Its closure semantics are configurable: either modulo or saturating. The
  fixpnt is constructed with blocks of type **BlockType**, and a fixpnt value is
  aligned in memory on **BlockType** boundaries.

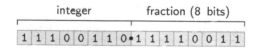

Fig. 3. A 16-bit fixed-point with 8 fraction bits.

cfloat<nbits, es, BlockType, sub, super, saturating> The **cfloat** arithmetic
  type is a floating-point type of size *nbits* bits, with 1 sign bit, an exponent
  field of *es* bits, and *nbits* − 1 − *es* number of mantissa bits. The exponent
  is encoded as a biased integer. The mantissa is encoded with a hidden bit
  for normal and supernormals numbers. Subnormal numbers are numbers
  with all exponent bits set to 0s. Supernormal numbers are numbers with all
  exponent bits set to 1s. Normal numbers are all encodings that are not subnormal or supernormal. Closure semantics can be saturating or clipping to

±∞. Figure 4 shows a 16-bit **cfloat** with 5 bits of exponent. When subnormals are selected and supernormals and saturating are deselected, this would represent a half-precision IEEE-754 FP16. If subnormals are deselected, it will represent the NVIDIA and AMD FP16 arithmetic type. Figure 5 shows a single-precision floating-point. The **cfloat** type can represent floating-point types ranging from 3 bits to thousands of bits, with or without subnormals, with or without supernormals, and with clipping or saturating closure semantics.

Fig. 4. Half-precision 16-bit floating-point representation, fp16.

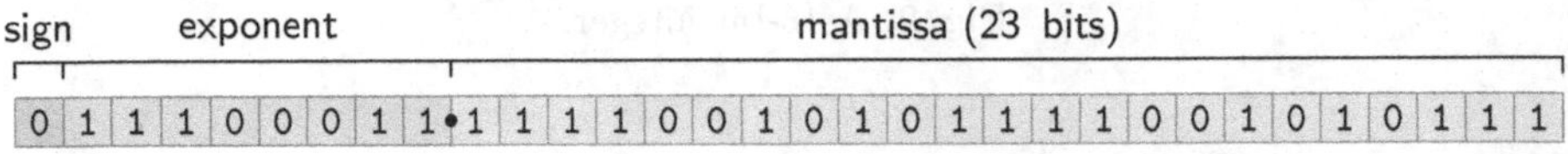

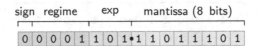

Fig. 5. Single precision 32-bit floating-point representation, fp32.

posit<nbits, es, BlockType> The **posit** arithmetic type represents a tapered floating-point type using the posit encoding. It offers a parameterized size of *nbits*, with 1 sign bit, *es* exponent bits, and *nbits* − 3 − *es* mantissa bits around 1.0. Figure 6 shows a 16-bit posit with 3 exponent bits in regime −3.

sign regime   exp   mantissa (8 bits)

| 0 | 0 | 0 | 0 | 1 | 1 | 0 | 1•1 | 1 | 0 | 1 | 1 | 1 | 0 | 1 |

Fig. 6. A posit<16,3>.

lns<nbits, base, BlockType> The **lns** arithmetic type implements a logarithmic number system of size *nbits*, with *base* as the base.

The type set **integer, fixpnt, cfloat, posit**, and **lns** provide a productive baseline of arithmetic types that most developers are familiar with. *Universal* contains other types as well, including faithful types with uncertainty bits, type I and II **unums**, and interval **posits**, called **valids**. These more advanced arithmetic types provide facilities for reliable computing and numerical analysis.

3.4   Variable Size, Adaptive Precision

Adaptive precision arithmetic types are tailored to questions regarding numerical precision and computational accuracy. This has been the traditional domain of numerical research platforms, such as Boost Multiprecision, MPFR, and GMP [5,7,19]. *Universal* currently offers only two adaptive precision arithmetic types:

- decimal
- rational

Implementation work has started on adaptive precision floating-point based on Douglas Priest's work [24], and the lazy exact arithmetic type proposed by Ryan McCleeary [20].

4   Creating a New Arithmetic Type

The experience with implementing a dozen or so arithmetic types has exposed a typical pattern of how to quickly and reliably bring up a new arithmetic type. The first step is to define the memory layout of the parameterized type. This step blocks the storage required to contain the encoded bits. *Universal* exposed the block type used for storage and alignment. For example, using a uint8_t as the building block, the memory layout of the individual value would be the minimum number of blocks to contain the encoding. The memory alignment would be on byte boundaries.

Once the memory layout has been designed, the next step is to implement the conversion from encoding to native types, such as **float** or **double**. Provide a simple set of convert_to_ methods to test the interpretation of bits in the encoding to generate tables to validate the encoding.

The next step is to implement the inverse transformation - the conversion from native types to the encoding of the new arithmetic type. This conversion tends to be the most involved algorithmic task as sampling the native type values by the new arithmetic type requires robust rounding decisions. The conversion regression suite of this step is also involved as one needs to enumerate all possible rounding situations across all possible encodings.

Once we have the memory layout, encoding, and the two conversion directions, the type can be used for computation by simply converting the value to a native type, calling arithmetic operators or math library functions, and converting the result back into the encoding.

The final two implementation tasks are native arithmetic operators and native implementations of the elementary functions. Native implementations are the only safeguard against double-conversion errors. Native implementations are also crucial for performance and hardware validation. The arithmetic operators vary by which algebraic system the arithmetic type is associated. Still, in general, we need to implement the following set:

- addition
- subtraction

- multiplication
- division
- remainder
- square root

The final step in creating a new arithmetic type is to provide native implementations of the elementary functions. To do this for parameterized types is still an open research question, as the approximation polynomials, albeit *minimax* or *Minefield*, are specific to each configuration.

5   Arithmetic Type API

5.1   Construction

All Universal arithmetic types have a default, copy, and move constructor. This allows the application to create, copy, and efficiently call return values.

```
// required constructors
constexpr posit() noexcept
constexpr posit(const posit&) noexcept
constexpr posit(posit&&) noexcept
```

To support efficient conversions between native types and the user defined type, we encourage to provide converting constructors for all native types. Casting to the largest native type would create inefficiency specifically for small encodings where performance is most desired.

```
// signed native integer types
constexpr posit(signed char) noexcept
constexpr posit(short) noexcept
constexpr posit(int) noexcept
constexpr posit(long) noexcept
constexpr posit(long long) noexcept

// unsigned native integer types
constexpr posit(char) noexcept
constexpr posit(unsigned short) noexcept
constexpr posit(unsigned int) noexcept
constexpr posit(unsigned long) noexcept
constexpr posit(unsigned long long) noexcept

// native floating-point types
constexpr posit(float) noexcept
constexpr posit(double) noexcept
constexpr posit(long double) noexcept
```

Some compilers, Clang in particular, treat type aliases as different types. Aliases such as uint8_t, uint16_t, uint32_t, uint64_t are not equivalent to char, unsigned short, unsigned int, and unsigned long long, respectively. This causes potential compilation problems when using type aliases in converting constructors. Instead of matching the appropriate constructor, your code will go through an implicit conversion, which can cause latent bugs that are hard to find. Best is to specialize on the native language types, short, int, long, etc.

5.2   Assignment

Assignment operators follow the same structure as the converting constructors.

```
// assignment operators for native types

// signed native integer types
constexpr fixpnt& operator=(signed char rhs)           noexcept
constexpr fixpnt& operator=(short rhs)                 noexcept
constexpr fixpnt& operator=(int rhs)                   noexcept
constexpr fixpnt& operator=(long rhs)                  noexcept
constexpr fixpnt& operator=(long long rhs)             noexcept

// unsigned native integer types
constexpr fixpnt& operator=(char rhs)                  noexcept
constexpr fixpnt& operator=(unsigned short rhs)        noexcept
constexpr fixpnt& operator=(unsigned int rhs)          noexcept
constexpr fixpnt& operator=(unsigned long rhs)         noexcept
constexpr fixpnt& operator=(unsigned long long rhs)    noexcept

// native floating-point types
constexpr fixpnt& operator=(float rhs)                 noexcept
constexpr fixpnt& operator=(double rhs)                noexcept
```

Another compiler environment constraint, particularly for embedded environments, is support for long double. Embedded ARM and RISC-V compiler environments do not support long double, so Universal guards the long double construction/conversion and must be explicitly enabled.

```
// guard long double support to enable
// ARM and RISC-V embedded environments
#if LONG_DOUBLE_SUPPORT
constexpr fixpnt(long double initial_value)            noexcept
constexpr fixpnt& operator=(long double rhs)           noexcept
constexpr explicit operator long double() const        noexcept
#endif
```

5.3   Conversion

Operators that convert from native types to the custom type are provided through converting constructors and assignment operators. However, the conversion from custom type to native types is marked explicit to avoid implicit conversions that can hide rounding errors that are impossible to isolate.

```
// make conversions to native types explicit
explicit operator int()          const
explicit operator long long()    const
explicit operator double()       const
explicit operator float()        const
explicit operator long double()  const
```

5.4   Logic Operators

The *Universal* arithmetic types are designed to be plug-in replacements for native types. Notably, for the logic operators in the language, it is common to come across this code:

```
...
cfloat<16,5> a, b, c;
...
if (b != 0) c = a / b;
...
```

Porting existing codes to use *Universal* types provided evidence that all and every combination of literal comparisons are used. The logic operator design must thus be complete and capture all combinations of arithmetic type and literal type that are native to the language.

The design in *Universal* uses a strongly typed set of operator signatures that provide an optimized implementation for the comparison leveraging the size of the literal. Typically, we only need to implement `operator==()` and `operator<()` with native encoding knowledge. The other logic operators can be expressed in terms of these two operators. The exception to this rule is the IEEE-754 derived arithmetic types with NaN encodings. In those systems, the logic operators are not complementary, and each operator requires decision code to deal with this particular type.

```
// base logic operators are defined as friends
template<size_t nbits, size_t es>
friend bool operator==(const valid<nbits, es>& lhs,
                       const valid<nbits, es>& rhs);
template<size_t nbits, size_t es>
friend bool operator!=(const valid<nbits, es>& lhs,
                       const valid<nbits, es>& rhs);
template<size_t nbits, size_t es>
friend bool operator< (const valid<nbits, es>& lhs,
                       const valid<nbits, es>& rhs);
template<size_t nbits, size_t es>
friend bool operator> (const valid<nbits, es>& lhs,
                       const valid<nbits, es>& rhs);
template<size_t nbits, size_t es>
friend bool operator<=(const valid<nbits, es>& lhs,
                       const valid<nbits, es>& rhs);
template<size_t nbits, size_t es>
friend bool operator>=(const valid<nbits, es>& lhs,
                       const valid<nbits, es>& rhs);
```

There are three signed integers (`int`, `long` and `long long`), three unsigned, and three floating-point (`float`, `double`, and `long double`) literals. This creates the need for $6 \times 3 \times 3 \times 3 \times 2 = 324$ free functions to capture all the combinations between arithmetic type and literal. These free functions will transform the literal to the arithmetic type and then call the operational layer to execute the comparison. Judicious use of implicit conversion rules can be used to reduce that number, but care must be taken to avoid double rounding errors.

5.5   Arithmetic Operators

Binary arithmetic operators are implemented through free binary functions that capture literals and type conversions coupled with in-place update class operators.

```
// update operators
cfloat& operator+=(const cfloat& rhs)
cfloat& operator-=(const cfloat& rhs)
cfloat& operator*=(const cfloat& rhs)
cfloat& operator/=(const cfloat& rhs)

// free binary functions to transform literals
template<size_t nbits, size_t es, typename bt,
bool hasSubnormals, bool hasSupernormals, bool isSaturating>
cfloat<nbits, es, bt,
       hasSubnormals, hasSupernormals, isSaturating>
operator+(const double lhs,
          const cfloat<nbits, es, bt,
                       hasSubnormals, hasSupernormals,
                       isSaturating>& rhs) {
        cfloat<nbits, es, bt, hasSubnormals, hasSupernormals,
               isSaturating> sum(lhs);
        sum += rhs;
        return sum;
}
```

5.6   Serialization

All arithmetic types in *Universal* support serialization through the stream libraries.

```
std::ostream& operator<<(std::ostream& ostr,
                         const decimal& d)
std::istream& operator>>(std::istream& istr, decimal& p)
```

Such conversions may introduce rounding errors, so *Universal* types also support a error free **ASCII** format. This is controlled by a compilation guard:

```
//////////////////
// enable/disable special posit format I/O
#if !defined(POSIT_ERROR_FREE_IO_FORMAT)
// default is to print double values
#define POSIT_ERROR_FREE_IO_FORMAT 0
#endif
```

When error-free printing is enabled, values are printed with a designation and a hex format to represent the bits. Here is an example of a posit use case:

```
posit<32,2> p(1.0);
cout << "Error free posit value: " << p << endl;
...
> Error free posit value: 32.2x40000000p
```

5.7   Set and Query Interface

To support efficient verification and validation of an arithmetic type, the regression suites need to be able to set and query bits.

```
// modifiers
inline constexpr void clear() noexcept
inline constexpr void setzero() noexcept
inline constexpr void setbit(size_t i, bool v = true)
inline constexpr void setbits(uint64_t raw_bits) noexcept
inline constexpr void setblock(size_t b, const BlockType& data) noexcept
```

The methods setbit() and setbits() make it possible to write generic regression tests for arithmetic types, thus drastically reducing the amount of code that needs to be written to validate the arithmetic types in *Universal*. This basic API can be augmented to set special encodings for specific arithmetic types. For example, here is the extended set for *cfloat*:

```
inline constexpr void setinf(bool sign = true) noexcept
inline constexpr void setnan(int NaNType = NAN_TYPE_SIGNALLING) noexcept
inline constexpr void setsign(bool sign = true)
inline constexpr bool setexponent(int scale)
inline constexpr void setfraction(uint64_t raw_bits)
```

The verification phase is aided by a productive reflection interface. The *Universal* arithmetic types have a standard set of state query methods to simplify the verification algorithms.

```
// selectors
inline constexpr bool sign() const noexcept
inline constexpr int  scale() const noexcept
inline constexpr bool ispos() const noexcept
inline constexpr bool iszero() const noexcept
inline constexpr bool isone() const noexcept
```

Just like the modifiers, the basic API can be augmented to capture specific state: here is the extended set for *cfloat*.

```
// special value queries for cfloat
inline constexpr bool isinf(int InfType) const noexcept
inline constexpr bool isnan(int NaNType) const noexcept

// range queries for cfloat
inline constexpr bool isnormal() const noexcept
inline constexpr bool isdenormal() const noexcept
inline constexpr bool issupernormal() const noexcept
```

5.8   Support Functions and Manipulators

Working with bit encodings is challenging, so *Universal* provides a collection of manipulators and support functions to ease queries and interpret bit encodings. The manipulator color_print color-codes different segments of encoding so that it is easier to decipher and compare, as is shown in Fig. 7 for different posit configurations.

```
posit< 8,0>   = 01101001 : 3.125
posit< 8,1>   = 01011001 : 3.125
posit< 8,2>   = 01001101 : 3.25
posit< 8,3>   = 01000110 : 3
posit<16,1>   = 0101100100100010 : 3.1416
posit<16,2>   = 0100110010010001 : 3.1416
posit<16,3>   = 0100011001001000 : 3.1406
posit<24,1>   = 010110010010000111111011 : 3.141592
posit<24,2>   = 010011001001000011111110 : 3.141594
posit<24,3>   = 010001100100100001111111 : 3.141594
posit<32,1>   = 01011001001000011111101101010100 : 3.14159265
posit<32,2>   = 01001100100100001111110110101010 : 3.14159265
posit<32,3>   = 01000110010010000111111011010101 : 3.14159265
posit<48,1>   = 010110010010000111111010101010001000100001101101 : 3.1415926535898
posit<48,2>   = 010011001001000011111101101010100010001000010111 : 3.1415926535899
posit<48,3>   = 010001100100100001111110101010101000100010000101011 : 3.1415926535897
posit<64,1>   = 0101100100100001111110101010100010001000010110100011000000000000 : 3.14159265358979312
posit<64,2>   = 0100110010010000111111011010101000100010000010110100011000000000 : 3.14159265358979312
posit<64,3>   = 0100011001001000011111110101010100010001000010110100011000000000 : 3.14159265358979312
posit<64,4>   = 0100001100100100001111110110101010001000100001011010001100000000 : 3.14159265358979312
```

Fig. 7. Demonstration of the `color_print(p)` function.

6   Example: Matrix Scaling and Equilibrating

Several multi-precision iterative-based algorithms for solving $Ax = b$ have been developed [2,3,8,10] showing speedup over double precision solvers (e.g., [11]). In those studies, the algorithms round the entries of A to lower precision (i.e., half precision), perform LU decomposition, compute a solution using the low accuracy LU factors, then use iterative refinement back to working precision. The results are highly dependent on the condition number of the matrix. It is worth noting that even well-conditioned matrices can become ill-conditioned in lower precision. Because when scaling to lower precision, underflow may produce a singular matrix, the percentage of nonzero elements after scaling to lower precision is critical. Rounding to lower precision also can result in overflow (or subnormals); however, overflow is far less likely in scientific computing applications [12]. Next, we outline three algorithms (see Appendix A) presented in [12] using slightly different notation.

Let p represent the precision (e.g., `fp16`), $a_{max} = \max_{i,j} |a_{i,j}|$, and x_{max} the largest positive value represented in the lower precision arithmetic. The goal of scaling is to reduce the condition number and to increase speed in solving. Table 1 lists key characteristics of half precision, double precision, and `posit< 16,2>`.

Algorithm 1 converts all entries to lower precision format. Any entry that rounds to infinity is replaced by the maximum signed value. However, Algorithm 1 does not handle underflow or subnormal situations. Furthermore, it alters the matrix significantly when $|a_{ij}| \gg x_{max}$, (see [12]). Algorithm 2 on the other hand, scales then rounds in such a way to avoid overflow, however shrinks the magnitude of each increasing the chance of underflow, thus increasing the possibility of producing a singular matrix. Algorithm 3 addresses those issues by equilibrating rows and columns so that the maximum entry is 1 in each row and column. We note it is not possible to underflow using posit configurations, so there will be no reduction in the number nonzero elements. As such, more research on how and when to scale is needed. Universal permits testing multiple configurations and provides a vehicle for this research.

Table 1. Specifications of multiple binary formats. Precision of the arithmetic is measured by the unit round off, which is 2^{-p}, where p is the number of fraction bits.

Format	Fraction	Exponent	ulp	Min	Max
posit<16,2>	≤ 11	2	4.88e−04	1.39e−17	7.21e+16
fp16	11	5	4.88e−04	6.10e−05	6.55e+04
fp64	53	11	1.11e−16	2.22e−308	1.80e+308

7   Conclusions and Future Work

We have presented the current status of the *Universal* library v3. The third edition of the library contains the core arithmetic types that cover the algebraic sets and contains a unified API to simplify validation and regression testing. We provided some examples of the arithmetic types' intended use and reported on some framework integrations for research and development of computational science and engineering applications. Much work remains. The Oracle-style arithmetic types need to be fleshed out so that the mathematical library work can move forward. Math libraries that are parameterized with arithmetic type attributes such as precision and dynamic range are still an open research question and, when resolved, would dramatically improve the code-efficiency of *Universal*. Many new arithmetic types are being proposed for Deep Learning and Optimal control that need representations that are more energy-efficient. Moreover, many next-generation application platforms are written in Python, Julia, Golang, and Rust, and will need integration facilities to leverage the arithmetic types in *Universal*. The current incarnation of the library provides productive facilities to research and develop energy-efficient and high-performance algorithms through custom arithmetic. The *Universal* library development is managed as an open-source community project [1] where all contributions are welcome.

Appendix A: Squeezing Algorithms

Algorithm 1: Round and replace overflow with x_{max}.

   Input: An $n \times n$ Matrix A
   Output: Rounded matrix B

1 $B = \mathrm{fl_p}(A)$
2 Set $a_{ij}^{p} = \mathrm{sign}(a_{ij})x_{\mathrm{max}}$

Algorithm 2: Scaled matrix entries

Input: An $n \times n$ Matrix A
Output: Rounded matrix $A^{(p)}$

1 $a_{\max} = \max_{i,j} |a_{i,j}|$
2 $\mu = x_{\max}/a_{\max}$
3 $A^{(p)} = \mathrm{fl_p}(\mu A)$

Algorithm 3: Double side scaling using row and column equilibration

---

Input: An $n \times n$ Matrix A
Output: Rounded matrix $A^{(p)}$

1 Set $R = 0$
2 **for** $i = 1$ **to** n **do**
3 $\quad R(i,i) \leftarrow \|A(i,:)\|_\infty^{-1}$
4 **end**
5 $B = RA$
6 Set $S = 0$
7 **for** $j = 1$ **to** n **do**
8 $\quad S(j,j) \leftarrow \|A(:,j)\|_\infty^{-1}$
9 **end**
10 Set $\beta =$ maximum absolute entry in RAS
11 Set $\mu = x_{\max}/\beta$
12 Return $\mathrm{fl_p}(\mu(RAS))$

References

1. Universal number library (2017). https://github.com/stillwater-sc/universal
2. Carson, E., Higham, N.J.: A new analysis of iterative refinement and its application to accurate solution of ill-conditioned sparse linear systems. SIAM J. Sci. Comput. **39**(6), A2834–A2856 (2017)
3. Carson, E., Higham, N.J.: Accelerating the solution of linear systems by iterative refinement in three precisions. SIAM J. Sci. Comput. **40**(2), A817–A847 (2018)
4. Cox, M.G., Hammarling, S.: Reliable Numerical Computation. Clarendon Press, Oxford (1990)
5. Fousse, L., Hanrot, G., Lefèvre, V., Pélissier, P., Zimmermann, P.: MPFR: a multiple-precision binary floating-point library with correct rounding. ACM Trans. Math. Softw. (TOMS) **33**(2), 13-es (2007)
6. Gottschling, P., Wise, D.S., Adams, M.D.: Representation-transparent matrix algorithms with scalable performance. In: Proceedings of the 21st Annual International Conference on Supercomputing, pp. 116–125 (2007)
7. Granlund, T.: GNU MP. The GNU Multiple Precision Arithmetic Library **2**(2) (1996)
8. Gupta, S., Agrawal, A., Gopalakrishnan, K., Narayanan, P.: Deep learning with limited numerical precision. In: International Conference on Machine Learning, pp. 1737–1746. PMLR (2015)
9. Gustafson, J.L., Yonemoto, I.T.: Beating floating point at its own game: posit arithmetic. Supercomput. Frontiers Innovations **4**(2), 71–86 (2017)

10. Haidar, A., Tomov, S., Dongarra, J., Higham, N.J.: Harnessing GPU tensor cores for fast FP16 arithmetic to speed up mixed-precision iterative refinement solvers. In: SC18: International Conference for High Performance Computing, Networking, Storage and Analysis, pp. 603–613. IEEE (2018)

11. Haidar, A., Wu, P., Tomov, S., Dongarra, J.: Investigating half precision arithmetic to accelerate dense linear system solvers. In: Proceedings of the 8th Workshop on Latest Advances in Scalable Algorithms for Large-Scale Systems, pp. 1–8 (2017)

12. Higham, N.J., Pranesh, S., Zounon, M.: Squeezing a matrix into half precision, with an application to solving linear systems. SIAM J. Sci. Comput. **41**(4), A2536–A2551 (2019)

13. Hittinger, J., et al.: Variable precision computing. Technical report, Lawrence Livermore National Lab. (LLNL), Livermore, CA (United States) (2019)

14. Horowitz, M.: 1.1 computing's energy problem (and what we can do about it). In: 2014 IEEE International Solid-State Circuits Conference Digest of Technical Papers (ISSCC), pp. 10–14. IEEE (2014)

15. Intel Corporation: BFLOAT16 - Hardware Numerics Definition (2018). https://www.intel.com/content/dam/develop/external/us/en/documents/bf16-hardware-numerics-definition-white-paper.pdf

16. Jouppi, N.P., et al.: In-Datacenter performance analysis of a tensor processing unit. In: Proceedings of the 44th Annual International Symposium on Computer Architecture, pp. 1–12 (2017)

17. Kharya, P.: TensorFloat-32 in the a100 GPU accelerates AI training HPC up to 20x. NVIDIA Corporation, Technical report (2020). https://blogs.nvidia.com/blog/2020/05/14/tensorfloat-32-precision-format/

18. Lloyd, G.S., Lindstrom, P.G.: ZFP hardware implementation. Technical report, Lawrence Livermore National Lab. (LLNL), Livermore, CA (United States) (2020)

19. Maddock, J., Kormanyos, C., et al.: Boost multiprecision (2018)

20. McCleeary, R.: Lazy exact real arithmetic using floating point operations (2019)

21. Molisch, A.F., et al.: Hybrid beamforming for massive MIMO: a survey. IEEE Commun. Mag. **55**(9), 134–141 (2017)

22. Muhammad, K., Ullah, A., Lloret, J., Del Ser, J., de Albuquerque, V.H.C.: Deep learning for safe autonomous driving: current challenges and future directions. IEEE Trans. Intell. Transp. Syst. **22**(7), 4316–4336 (2020)

23. Omtzigt, E.T.L., Gottschling, P., Seligman, M., Zorn, W.: Universal numbers library: design and implementation of a high-performance reproducible number systems library. arXiv:2012.11011 (2020). https://arxiv.org/abs/2012.11011

24. Priest, D.M.: Algorithms for Arbitrary Precision Floating Point Arithmetic. University of California, Berkeley (1991)

25. Siek, J.G., Lumsdaine, A.: The matrix template library: a generic programming approach to high performance numerical linear algebra. In: Caromel, D., Oldehoeft, R.R., Tholburn, M. (eds.) ISCOPE 1998. LNCS, vol. 1505, pp. 59–70. Springer, Heidelberg (1998). https://doi.org/10.1007/3-540-49372-7_6

26. Siek, J.G., Lumsdaine, A.: The matrix template library: a unifying framework for numerical linear algebra. In: Demeyer, S., Bosch, J. (eds.) ECOOP 1998. LNCS, vol. 1543, pp. 466–467. Springer, Heidelberg (1998). https://doi.org/10.1007/3-540-49255-0_152

27. Dally, W.J., et al.: Neural network accelerator using logarithmic-based arithmetic (2021). https://uspto.report/patent/app/20210056397

Small Reals Representations for Deep Learning at the Edge: A Comparison

Marco Cococcioni[1] , Federico Rossi[1]([✉]) , Emanuele Ruffaldi[2] ,
and Sergio Saponara[1]

[1] Università di Pisa, Pisa, Italy
{marco.cococcioni,federico.rossi,sergio.saponara}@unipi.it
[2] MMI s.p.a, Calci, Italy
emanuele.ruffaldi@mmimicro.com

Abstract. The pervasiveness of deep neural networks (DNNs) in edge devices enforces new requirements on information representation. Low precision formats from 16 bits down to 1 or 2 bits have been proposed in the last years. In this paper we aim to illustrate a general view of the possible approaches of optimizing neural networks for DNNs at the edge. In particular we focused on these key points: i) limited non-volatile storage ii) limited volatile memory iii) limited computational power. Furthermore we explored the state-of-the-art of alternative representations for real numbers comparing their performance in recognition and detection tasks, in terms of accuracy and inference time. Finally we present our results using posits in several neural networks and datasets, showing the small accuracy degradation between 32-bit floats and 16-bit (or even 8-bit) posits, comparing the results also against the bfloat family.

Keywords: Deep learning · Edge computing · Fog computing · Fine tuning at the edge · Alternative representation for real numbers · Small reals · Posits · bfloat · Weights compression

1 Introduction

Recently, it has been shown that Machine Learning in general, and Deep Neural Networks (DNNs) in particular, tolerate low-precision representations for their parameters. This constitutes an opportunity for speeding up the computations, to reduce storage, and, more importantly, to reduce power consumption. The latter is of paramount importance at the edge and on embedded devices.

However, to allow the porting of trained DNNs on difference devices, there is the need to standardize low precision formats for machine learning.

The aim of this work is to grab the attention to this very important topic, with the hope that sooner or later a standard, like the well-known IEEE 754 one (see [1]), will be put in place.

This is a necessity strongly felt by practitioners and industry, even if academics and researchers seem to be less aware of its importance.

© The Author(s), under exclusive license to Springer Nature Switzerland AG 2022
J. Gustafson and V. Dimitrov (Eds.): CoNGA 2022, LNCS 13253, pp. 117–133, 2022.
https://doi.org/10.1007/978-3-031-09779-9_8

To make the picture of the situation more complex, we should also take into account the requirements of safety critical applications, where low-precision is less encouraged, but can still be considered, provided that it does not hamper the safety of the system.

Safety critical applications at the edge not only put more stringent requirements on the binary representation for small reals in DNNs, but can also add constraints of reproducibility of the computations. This latter aspect can impact the design of the representation. As an example, consider the use of stochastic rounding: even if it has been proved to increase the effectiveness during the training of a DNN (especially when using 8-bit precision floating point numbers [2]), it undermines the reproducibility of the computations. Since we are confident that sooner or later a standard will be created, it is important to start to make comparison between the existing alternative ways to represents real numbers in deep neural networks, in particular when planned to be used at the edge. Before doing this, we provide a review of all the techniques proposed so far to reduce power consumption, such as quantization, network pruning, etc.

The paper is organized as follows: in Sect. 2 we reviewed the state of art of deploying Deep Neural Networks at the edge and the main trends of research activities in this field. In Sect. 2.1 we briefly described the network pruning technique and its applications in simplifying neural networks. In Sect. 2.2 we summarized the network quantization approach, also covering networks working with binary or ternary weights (we have called the latter cases as "drastic quantization"). In Sect. 2.3 we reviewed a family of low-precision format for DNNs, called small reals, that include all the types we analysed later on. In Sect. 3 we analysed the most promising alternatives to IEEE 32-bit floats: bfloat family in Sect. 3.1, flexpoint in Sect. 3.2 and Logarithmic Numbers in Sect. 3.3. In Sect. 4 we presented and deeply analysed the posit^{TM} format, highlighting some important properties. Furthermore, we showed the main contributions of this work, consisting of the integration of the cppPosit library and bfloats inside some interesting machine learning frameworks. In Sect. 5 we presented results on deploying neural networks on a low-power constrained device, the Raspberry Pi 3B and in Sect. 6 we analyse the obtained results and their impact, other than discussing future developments of the proposed approach. Finally, in Sect. 7 we draw a few conclusions.

2   Deploying DNNs at the Edge: State of the Art

In the last decade, a lot of research efforts in DNNs has been devoted to reduce the resources required to exploit neural networks with limited memory, storage or computing power (such as smartphones or network edge devices), as demonstrated by the success of TensorFlow Lite, the low-precision counterpart of Google TensorFlow library. Two research lines emerged, the first one focusing on the inference phase only, leading to reduced-precision representation for the neural network parameters, the second one aimed at additionally speeding up the training phase using low-precision numerical formats also for the gradients.

Concerning low-precision numerical formats currently used in DNNs, three main approaches can be distinguished:

1. use of low-precision floating-point formats;
2. use of low-precision fixed-point real numbers or integer numbers;
3. use of binary/ternary formats.

These alternative representations can be limited to the weights, or to the weights and activations, or include all involved quantities (weights, activations and gradients). When following the first approach (i.e., low-precision floats), research and development efforts are converging toward a 16-bit floating point representation instead of the classical 32-bit one (which is called binary32) [1]. The same IEEE 754 standard [1] which has standardized binary32 has also standardized a 16-bit counterpart, called binary16, which reserves 5 bits to the exponent. However, most of the general purpose CPUs do not have full hardware support for binary16. In addition, it seems to be not particularly effective in deep learning. For these reasons, IBM has proposed a 16-bit floating point format having 6 bit for the exponent [3], called DLFloat (which stands for "deep learning float"), while Google has proposed the 8-bit alternative for the exponent [4], called bfloat16. This gap in the standard might be resolved soon, as there is a strong push from the machine learning community for suitable arithmetic formats. Another shortcoming of this approach is the lack of hardware support: as said above, most CPUs support 32- and 64-bit floats, but not 16- or 8-bit floats. Moreover, there are proposals to use a completely different representation for real numbers, like the posit format introduced in 2017 [5]. Although the posit format is promising for low-precision DNNs [6–9], the not widespread availability of hardware support on CPUs still limits a large scale adoption (a list of hardware implementations of posits can be found at https://en.wikipedia.org/wiki/Unum_(number_format)).

The second approach (i.e., low-precision fixed-point numbers or integer numbers) is popular since it allows running DNNs even on entry-level CPUs microcontrollers not equipped with a Floating Point Unit (FPU), since the Arithmetic Logic Unit (ALU) is enough. On the one hand, fixed-point representations for real numbers are widely used (especially in financial applications and to improve the graphics in video games) even though C++ has not yet a standard library supporting them. DNN implementations using low-precision fixed-point for the both the weights and the activations are appearing [10]. Recently, a few papers discussed the specific issues of training DNNs with a fixed-point representation [11]. On the other hand, low-precision integer numbers are very interesting for time-sensitive applications, because operations between integer numbers have predictable computing times. However, the use of (low-precision integers), like 8-bit or less, usually requires a tailored training algorithm [12]. This approach is called *quantization*, for its obvious meaning.

The third approach takes the use of low-precision integer numbers in DNNs to the extreme, using ternary or even binary weights. Remarkable results have been obtained: DNNs with ternary weights (i.e., -1, 0 and 1) have been demonstrated to achieve the same classification accuracy as DNNs using binary32 weights [13]. DNNs with binary weights have been also devised, again with little or no degradation in the classification accuracy [14]. These results were confirmed on the very challenging ImageNet dataset, considered as the most demanding open-source dataset for visual object recognition, with more than 20,000 different object categories [15]. The use of models with precision down to INT2 (i.e., 2-bit integer) has been demonstrated with a more than tolerable accuracy loss [16,17]. As a result, NVIDIA has added the support down to binary numbers to its top-level GPUs to perform tensor operations [18]. Quantization can be applied either during the training phase or after it, just to perform the inference. However, DNN training using these numerical formats is more difficult compared to the two previously presented solutions as the gradient descent cannot be exploited, requiring the implementation of ad-hoc learning algorithms.

In [19] a series of challenges for DNN edge computing was presented. In particular the authors pointed out 4 main challenges to obtain a so called "TinyML": i) profiling the energy consumption is critical and the power consumption can vary wildly between different devices ii) edge systems often have very limited memory, two orders of magnitude smaller than usual smartphones, so optimizations are required iii) edge devices can be very different from each other, thus there is the need to normalize the benchmarks and the results obtained in those heterogeneous environments. iv) there is the need for software abstraction, even if this means losing a bit of low-level optimization that comes from hand-written and hand-tuned code.

2.1 Network Pruning

When deploying trained model to edge devices we must balance the model accuracy performance with the inference processing time and resource utilisation. Indeed, the principal aim of network pruning is to reduce the computational cost of DNNs.

Typically DNNs are deployed with a large number of layers if several types, with most of them having their own weights and feature maps: traditionally, pruning is aimed to drastically reduce the amount of parameters in the network by removing some "redundant" connection between the layers. The idea is to delete such parameters whose removal will impact the less the training error. For example, we can delete very small-magnitude weights (when compared to the rest of the network). After deletion the network can continue its training, and so on, deleting weights at each step applying different deletion strategy. As a drawback the training process is significantly slowed down, since it requires a particular fine-tuning after each pruning step. The core idea expressed in [20] is to express an *optimal brain damage*, that is a theoretical measure of the *saliency* of weights in a network. In particular, a model of the training error functions is built and it is analytically associated with the effect of a perturbation of the parameters. From this expression the authors can express, with a series of transformations, the *saliency* s_k for each parameter k in the network.

The paper iterative approach is explained hereafter:

1. The neural network is first trained until a result (good enough) is reached
2. Each parameter is associated with a *saliency*, s_k
3. The parameters are sorted according to s_k and low-saliency parameters are deleted. Go to (1).

Deleting a parameter means setting it to 0 and making it immutable from that moment on.

2.2 Network Quantization

During the years, as the deep neural networks models grew in accuracy over the most famous datasets (e.g. ImageNet and others) the network complexity in terms of Floating Point operations (FLOPs) and model footprint increased. In particular the footprint of network models (e.g. AlexNet model size is around 233 MB) is particularly critical in low-power and edge devices that can be particularly constrained in non-volatile storage capacity. Typically quantization involves the compression of weights using small integers, like 8-bit integer types.

In [21] the authors presented an overview of quantization techniques on deep neural networks. In particular the authors were able to compress complex networks like AlexNet by a factor 35, using a combination of quantization and weight sharing, while inducing a very minimal increase in the recognition error.

Drastic Quantization (Binary and Ternary Networks). Another approach to quantization is pushing the compression further, aiming to binary or ternary weights representations. In [22] the authors presented an overview of several approaches to drastic quantization, using the Hybrid Binary Neural Network (HBN) model. This model is based on a combination of 32-bit integer layers and binary layers. Typically, the input layer and the prediction output layer have 32-bit integer weights, while the intermediate ones are implemented using binary weights. We report an example of HBN from Quantized Keras (QKeras), where the 95% of weights are binary:

1. 2-dimensional convolution with 32-bit weights
2. Batch Normalization with 32-bit weights
3. Quantization layer
4. 2-dimensional convolution with binary weights
5. Batch Normalization with 32-bit parameters
6. Quantization layer
7. Fully Connected layer with binary weights
8. Output layer with 32-bit predictions.

The authors showed how the choice of layers to be quantised (binarised in this case) is critical to reduce the network footprint and complexity without impacting the accuracy of the model.

2.3   Small Reals

Since quantization employs vary small integers for numerical representation, we lose the possibility to fine-tune our models on the edge without changing any aspect of the training algorithm. The idea of using *small reals* is based on the need for continuity between the original network model representation and the edge one. In particular, we want to remain in the real number domain. There are several formats that can be classified as small reals, each of them having at most 16-bit representation:

1. binary16: the standard IEEE 16-bit representation with 5-bit exponent and 10-bit fraction
2. bfloat family (in detail in Sect. 3.1) with 16 or 8 bit representations
3. posit numbers (in detail in Sect. 4) with 16 or 8 bit representations (but also intermediate variants can be used, such as 6, 10, 12 or 14 bits if we accept the cost of memory misalignment).

In the next section we provide the state of the art for alternative representations of real numbers, with special emphasis on small ones (16-bit or less). Then, in Sect. 4, we review the posit format, which is considered at the moment the main challenger to the IEEE 754 format.

3   Alternative Real Number Representation: State of the Art

3.1   The bfloat Family

The bfloat family is an alternative representation to IEEE 32-bit floating point numbers. In particular, the aim of bfloat is to propose a format that has very common characteristics with the IEEE 32-bit format, with a reduction on the format length.

bfloat16. The first format proposed in this family was the bfloat16. We summarize hereafter its structure:

- 1-bit sign
- 8-bit exponent
- 7-bit fraction

It substantially differs from its predecessor 16-bit IEEE Floating Point (binary16) because it has the same number of exponent bits as the 32-bit IEEE Floating Point (binary32). This allows a very fast conversion between the two types, since it only involves a truncation on the fraction (or an appropriate rounding, depending on the cases). This format can be employed both for low-precision inference and for mixed precision training [23].

There is a light support for bfloat16 in the latest generations of Intel Xeon CPUs; in particular BF16 instructions were added to the AVX2 vector extension of the architecture.

bfloat8. The bfloat8 format represents a further reduction in bits. Indeed, the format employs 5-bit exponent (as binary16) and only 2-bit for the fraction. This choice makes the conversion from binary16 to bfloat8 very fast, being it just a matter of truncation. The same cannot be said from binary32: in this case the conversion is more complex. A particular implementation of bfloat8 (in [24]) enabled the use of stochastic rounding during mixed precision training on this format. This approach allowed bringing in more randomization into the training phase. Let us consider a number represented on a float with a higher number of bits, let us say k bits, and we want to find its representation on k' bits, with $k' < k$. Let $x = s \cdot 2^e \cdot (1 + f)$ (sign, exponent and fraction respectively) be such a number. As an example, x might be a binary32 and x' a bfloat8. We may compute the probability $p = \frac{f - f'}{\epsilon}$ where f' is the truncation of the larger fraction into the smaller one and $\epsilon = 2^{-k}$. With probability p we round x to $y = s \cdot 2^e \cdot (1 + f' + \epsilon)$, while with probability $1 - p$ we round it to $y = s \cdot 2^e \cdot (1 + f')$. With this approach the authors were able to train several neural networks model on common datasets (e.g. CIFAR10 and ImageNet) with 8-bit floating point numbers: they reported very little degradation in DNN accuracy while reducing the model size by a factor ~ 2.

3.2 Flexpoint

The flexpoint format [25] is characterized by a *shared tensor exponent*. This exponent is used as a common exponent for all the reals in a given neural network layer or slice. This allows for example to have a 16-bit fixed-point representation in an entire DNN layer, with just additional 5-bits for the whole layer as an exponent. The exponent can be adjusted during the training, to match dynamic range variations that happen during the process. It should be noted that the idea behind flexpoint was already introduced earlier, but in different contexts, as "block floating point" representations (see, for example, [26]). Finally observe how the flexpoint approach, although interesting and powerful, cannot be used as a drop-in replacement to binary32: software changes are required to the DNN software libraries. This also makes cumbersome the reuse of pre-trained DNNs.

3.3 Logarithmic Numbers

As reported in [27], the main problem with floating point operations in hardware is the transistors occupation for multiplication and division operations, which occupy the main part of the FPU, being significantly more complex than the circuitry for addition/subtraction. To address this issue, the Logarithmic Number System (LNS) was proposed decades ago in [28]. This system represents a number x a number as $y = 2^x$, in a pure logarithmic way. Following the logarithm properties this means that multiplication and division are just a matter of adding and subtracting logarithmic numbers (e.g. $y_1 \times y_2 = 2^{x_1 + x_2}$).

However, this approach requires huge hardware lookup tables to compute the sum or difference of two logarithmic numbers [27]. This has been one of the main bottlenecks for the format, since handling these tables can be more expensive than basic hardware multipliers.

Although this approach is really promising and can be combined with other formats, it has not been demonstrated yet that logarithmic numbers are more effective than floats for DNNs. Thus more research is clearly needed before resorting to this solution.

4   The Posit Format and Innovative Contributions of This Work

Posit numbers [5] are a representation for real numbers that can be configured in two parameters, the number of bits *nbits* and the maximum number of exponent bits *esbits*.

The format can have at most 4 fields (3 when *esbit* is chosen equal to 0):

1. 1-bit sign
2. Variable length *regime*
3. Variable length (up to *esbits* if present) exponent[1]
4. Variable length fraction[2]

The novelty of the format is all in the regime field. This field is encoded with a run-length approach; indeed, its value depends on its length. To compute the length l of the regime we just need to look at the number of subsequent identical bits, interrupted by a bit of the opposite value (e.g. the bitstring 1110 has a length of 3, as well as the bitstring 0001). To compute the actual value of the regime we need to take into account also the value of the single bit that is repeated in the sequence, let's call it b. The regime value is then:

$$k = \begin{cases} -l, \text{ if } b = 0 \\ l - 1, \text{ otherwise} \end{cases} \quad (1)$$

The strong point of the variable length format is inside the regime: when numbers are small (around the values ±1), the regime length is low and the fraction length is high, thus giving the numbers in this area a high decimal accuracy. This makes perfectly sense when matched with Deep Neural Networks, where we can keep weights and activations across the layers near ±1 exploiting weight decay and normalization techniques. Furthermore, if we look at the posit range, most of the values are in the range $[-1, 1]$; this means that, a neural network whose weights are entirely contained in this range will lose very little accuracy if represented using posit numbers [29].

Particular properties of the posit format emerge when configuring the format with 0 exponent bits. In detail:

---

[1] An different way to look at the exponent field is to consider it having a fixed length of *esbit*, where possible missing ones bits are implicitly considered equal to zero.

[2] The same consideration done for the exponent field also applies to the fraction, which could be regarded as a fixed-length field too, with implicit zeros.

1. In the range $[-1, 1]$ it is identical to a fixed point format
2. Simple operations such as doubling, halving and inverting can be computed without decoding, directly on the posit integer representation [29]
3. Several DNN activation functions can be computed decoding free (see next section)

4.1   Fast Approximated Activation Functions

When we configure posit numbers with 0 exponent bits we can implement DNN activation functions using fast and approximated versions that can be computed directly on the integer representation, without decoding the posit.

Fast Approximated Sigmoid. As pointed out in the original posit paper, the Sigmoid can be computed directly on the posit representation as follows (v is the integer representing the argument number, while Y is the integer representing the result number):

$$Y = ((1 \ll nbits - 1) + v + 2) \gg 2$$

Fast Approximated Hyperbolic Tangent. From the sigmoid function we can build other activation functions by manipulating the expression using the operations described in the previous section (doubling, halving, inverting and others). The hyperbolic tangent (tanh) can be implemented using the following expression (if substituting the sigmoid with its approximated version, we obtain the fast approximated tanh):

$$tanh(x) = -(1 - 2 \cdot \text{sigmoid}(2x))$$

Fast Approximated Extended Linear Unit. The same approach can be followed with the Extended Linear Unit (ELU), by combining the fast approximated sigmoid function and the other approximated operations seen above:

$$e^x - 1 = -2 \cdot \left[1 - \frac{1}{2 \cdot \text{sigmoid}(-x)} \right]$$

In [30] the authors proposed a way to adopt posit numbers at the edge. They introduce a variant of posits called adaptive posit weight representation. When converting weights from 32-bit float representation they are also quantised to the adaptive posit format. This posit variant has a hyperparameter that control the dynamic range; it can be defined as a regime bias or as a maximum regime bit-width called rs. When $rs = 1$ the adaptive posit format is identical to a floating point with the same number of bits (the regime is non-existent in this case). When $rs = n - 1$ the adaptive posit format is a pure posit number. Thanks to this approach the authors were able to test their approach on different datasets and neural networks, without losing too much accuracy even with 5-bit adaptive posits. Furthermore they reported the maximum frequency (on

ASIC) obtained during conversions, peaking 1200 MHz with pure posit to float conversion with 5-bit posits. On the contrary, in this work we used standard 16-bit and 8-bit precision posits, and we have compared them with bfloat16 and bfloat8, respectively. The results of this comparison are reported in Sect. 5.

The aim of our approach is to compare different representations of real numbers on DNN fine tuning at the edge, *avoiding any change in the training algorithm*. In particular, we replace binary32 with bfloat16, bfloat8, posit16 and Posit8, and we report their classification accuracy on standard DNN classification benchmarks.

The added-value of this approach is that no software-hardware change are required, other than having an FPU supporting bfloat16/8 or posit16/8.

In particular, we are not requesting the support of the Stochastic Rounding, nor a different loss function or a tailored training algorithm.

In order to support posit numbers, in the past, we developed the cppPosit library [31]. To us, declaring a posit number is just simple as:

```
auto p8 = Posit<8,0,...>;
```

The greatest struggle in designing such a library was that we wanted a format that could be plug and play, so that we could just add it to any other machine learning library with just a type definition. To achieve this goal we focused on some core aspects of *modern* C++ (from 11 to 17):

- Type traits
- Extensive use of *constexpr* keyword to evaluate most of the branches at compile time, to gain as branchless portions of code as possible
- Extensive use of templates to generalize posit operations when compiling the code using `-Ofast -std=c++17`

When using novel types such as posit numbers, the lack of hardware is a critical aspect. We explicitly did not want to compute operations on posits (e.g. addition/multiplication and other) directly manipulating the posit bits. Instead, we only wrote the coding and decoding operations and the conversions to another type, called *backend*. The backend is a type that can leverage hardware acceleration to some extent. For example, two widely used backends in cppPosit are the fixed-point backend and the floating-point backend. Moreover, using a look-up table as a backend for such operations proved to be effective, but at greater memory cost.

Another obstacle to seamless integration of cppPosit with machine learning libraries was the interoperability with standard math library `<cmath>` or other linear algebra libraries (e.g. Eigen). Thanks to the extensive use of templates we easily integrated these two libraries within the cppPosit library, so that it could be easily used both with Eigen and the standard C++ math library.

This kind of interoperability out-of-the-box is not common in other posit libraries such as SoftPosit, that leverages a SoftFloat-like approach to arithmetic emulation. Furthermore, the cppPosit library is header only, therefore, its integration in a machine learning framework is simplified to just the inclusion of the main posit.h header.

Thanks to these design choices we integrated the posit library into the following machine learning framework:

- tinyDNN [32]: CPU-oriented DNN framework for small neural networks
- TensorFlow [33]: one of the most used DNN libraries, which offers a huge collection of datasets and pre-trained models.

A particular mention must be done to our posit-based TensorFlow implementation:

- The posit format was integrated *alongside* the other formats as a new *dtype* (a dtype is a data format in the TensorFlow name scheme)
- We needed to write a Python wrapper for cppPosit to accommodate the high-level Python interface.

As a result, we could load, store and convert pre-trained networks between posit format and the other format available in the TensorFlow library. In particular, we could manage to use 8-bit posits in TensorFlow (that typically does not allow 8-bit formats outside Tensorflow Lite) *without passing through network optimization and quantization* that are applied to the other 8-bit formats in TensorFlow. We achieved this by leveraging the posit encapsulation to mask the 8-bit type with a 16-bit memory alignment.

5   Comparison Results

In this section we present some results on deploying neural networks on a constrained resource device. We used a Raspberry Pi 3B, equipped with a Cortex-A53 (ARMv8) CPU running at 1.4 GHz and 1 GByte of LPDDR2 SDRAM. We tested neural network models that were trained in a much more powerful system using 32-bit floating point numbers. Then we converted such models to different numerical formats to evaluate the accuracy degradation of such representation. Furthermore, we reported the sample inference time of the models on the Raspberry board.

We used the following network models:

- LeNet-5 like convolutional neural network [34],
- EfficientNet deep convolutional neural network [35]
- Single Shot Detector (with 300×300 input images) [36]

We used the following evaluation datasets:

- MNIST: hand-written digits recognition benchmark [37], 32×32 grey scale images
- GTSRB: German Traffic Sign Recognition benchmark [38], 32×32 RGB images

- CIFAR10: general purpose image recognition benchmark [39], 32 × 32 RGB images
- ImagenetV2: additional test-set that uses the same Imagenet classes but with new images [40], 224 × 224 RGB images
- Pascal VOC 2007: object detection dataset [41], 300 × 300 RGB images

In Table 1 we reported the accuracy results of the first three small datasets (MNIST, GTSRB, CIFAR10) with the LeNet-5 like neural network. Since we hand-trained on these three datasets, we were able to add a normalization on our data pipeline, in order to represent the images on the range $[-1, 1]$, enabling us to perform inference using low-bit posits and bfloat.

In Table 2 we reported the accuracy results of the big datasets (Imagenet, PASCAL VOC) with the very deep neural networks (EfficientNet and SSD300). Since we were using a pre-trained model, we could not control the image encoding; indeed, the images in these two model were expected to be encoded in $[0, 255]$. This prevented us to use 8-bit posits and 8-bit bfloat due to numerical ranges.

Table 1. Inference (test-set) accuracy on small, edge convolutional neural network trained with binary32, on different small datasets.

	LeNet-5 like CNN		
	MNIST	GTRSB	CIFAR10
binary32	98.86%	91.9%	83.5%
posit16,1	98.83%	91.8%	83.5%
posit16,0	98.50%	90.5%	83%
bfloat16	98.86%	91.9%	82%
posit8,0	98.34%	90.4%	78%
bfloat8	69.57%	80.45%	67.5%

Table 2. Inference accuracy test on very deep neural networks with big datasets (again, pre-trained using binary32).

	EfficientNetB0 + ImagenetV2 (accuracy)	SSD300 + VOC 2007 (mean avg. precision)
binary32	81.9%	80.39%
posit16,2	79.7%	78.49%
bfloat16	78.9%	73.29%

Table 3. Sample inference time (frames per second in brackets) on different neural network models and input size. The times were evaluated on a Raspberry Pi3 Model B. Concerning posit16,x and posit8,x, we used $x = 0, 1, 2$ exponent bits, without observing changes in the speed.

	LeNet-5		EfficientNet	SSD300
Input Size:	$32 \times 32 \times 1$	$32 \times 32 \times 3$	$224 \times 224 \times 3$	$300 \times 300 \times 3$
posit16,x	9.2 ms (108.5 fps)	23.9 ms (41.72 fps)	17.05 s (0.05 fps)	730 s (0.0010 fps)
bfloat16	4.8 ms (208.97 fps)	9.7 ms (103.37 fps)	12.73 s (0.08 fps)	472 s (0.0020 fps)
posit8,x	9.1 ms (110.38 fps)	21 ms (46.94 fps)	15.89 s (0,06 fps)	714 s (0.0013 fps)
bfloat8	5.7 ms (173.03 fps)	11 ms (86.11 fps)	11.49 s (0.09 fps)	528 s (0.0018 fps)

6   Analysis of the Results and Future Developments

In Table 1 we can see how different formats perform in a scenario with small networks and simple datasets. As reported, all the 16-bit alternatives we analysed matched the baseline accuracy of the IEEE 32-bit floating point format. If we halve the bits again, with the 8-bit formats, we can see how 8-bit posits widely outperform bfloat8 numbers. This result show how 8-bit posits benefits from the non-fixed fraction bits, having the possibility to expand them at the expense of the regime when numbers are small. On the other hand, having only 2-bit of fraction in bfloat8 can be an issue when we plug directly the novel format without fine-tuning; indeed, if we could fine-tune the networks for a few epochs using only bfloat8 we could benefit from the stochastic gradient approach. However, without bfloat8 proper hardware support, this approach is still not feasible due to emulation overhead. We could think of applying the same idea to posit numbers, adding the support for such characteristic to a possible Posit Processing Unit (PPU).

From Table 2 we can see the behaviour of the 16-bit formats, when employed in more complex neural networks (EfficientNet has around 800 layers) and more challenging DNN tasks. As reported, the two 16-bit formats struggle to match the baseline accuracy, with the posit format losing 2% point in both cases and the bfloat16 losing respectively, 3 and 7% points.

In Table 3 we can see the sample inference time of the various networks, with different input sizes. When analysing these results we need to take in mind that we are completely emulating the behaviour of the different formats since we clearly do not have the proper hardware support and acceleration. Indeed, we rely on a floating point backend to perform the computation while weights and images are stored using the emulated format. This means that, for each multiplication or addition, we will convert the number to the floating point backend and then we will convert it back after computation to the original emulated format. This results show how bfloat family largely benefits from the strict similarity with IEEE floats; indeed, the conversion between a bfloat16 number and a binary32 one, is just a left shift of 16 positions (and vice versa) while the conversion between a posit and a binary32 numbers is way more complex, involving more operations.

If we combine both results from the tables we can envision a scenario where we use a 16-bit bfloat16 to perform mixed precision inference/training on 16-bit while we stick to posit8 for low-precision inference, having a clear advantage over bfloat8 in our tests.

6.1 Future Developments

When analysing bfloat8 we saw that it could benefit from a few epochs of fine-tuning using the stochastic rounding proposed by the authors. The most common framework that employs 8-bit formats, such as Tensorflow/Lite, widely use the quantization technique and network pruning to simplify networks for deployment in edge devices. This approach can introduce some issues: i) loss of performance in terms of accuracy ii) no guarantee of meeting target platform requirements iii) no guarantee on inference time or frames processed per second. An idea could be optimizing the network adding a multi-objective genetic algorithm that takes into account some parameters as constraints to match the target platform: i) maximum number of hidden layers, and ii) maximum number of active neurons per layer. With such constraints, we will be able to control both the time complexity for the training, the RAM request, and the inference latency (which, on his turn, impacts the frame per second that can be processed, in computer vision applications).

Future works may involve exploiting posit numbers for a family of micro-controllers that are equipped with an FPU (e.g. STM32 or Cortex F4) to be used as back-end unit for the computation.

7   Conclusions

In this paper we reviewed several techniques to optimize neural network for deployment to the edge. We have highlighted the quest for a new standard for computations with small reals at the edge. In particular we analysed the behaviour of two very promising formats, the bfloat family and the posit format. We presented some results concerning the use of the posit representation and compared them to results with bfloat numbers. From the results we saw that 16-bit posits and bfloat can match the baseline IEEE 32-bit float accuracy in several DNN task. Furthermore, we saw how 8-bit posit can outperform 8-bit bfloat in simple DNN tasks. Despite the good results obtained so far using posits, we think that there is still much to explore in order to fully exploit the potential of this novel format. In particular we expect to obtain more interesting results with the proper hardware support for both posit numbers and bfloat, which would allow the native training of really deep neural networks, or the fine tuning at the edge.

Acknowledgments. Work partially supported by H2020 projects (EPI grant no. 826647, https://www.european-processor-initiative.eu/ and TEXTAROSSA grant no. 956831, https://textarossa.eu/) and partially by the Italian Ministry of Education and Research (MIUR) in the framework of the CrossLab project (Departments of Excellence).

References

1. IEEE standard for floating-point arithmetic. IEEE Std 754-2019 (Revision of IEEE 754-2008), pp. 1–84 (2019). https://doi.org/10.1109/IEEESTD.2019.8766229
2. Mellempudi, N., et al.: Mixed precision training with 8-bit floating point (2019). arXiv: 1905.12334 [cs.LG]
3. Agrawal, A., et al.: DLFloat: a 16-b floating point format designed for deep learning training and inference. In: 2019 IEEE 26th Symposium on Computer Arithmetic (ARITH'19), pp. 92–95 (2019). https://doi.org/10.1109/ARITH.2019.00023
4. Burgess, N., et al.: Bfloat16 processing for neural networks. In: 2019 IEEE 26th Symposium on Computer Arithmetic (ARITH), pp. 88–91, June 2019. https://doi.org/10.1109/ARITH.2019.00022
5. Gustafson, J.L., Yonemoto, I.T.: Beating floating point at its own game: posit arithmetic. Supercomput. Frontiers Innovations 4(2), 71–86 (2017)
6. Cococcioni, M., Ruffaldi, E., Saponara, S.: Exploiting posit arithmetic for deep neural networks in autonomous driving applications. In: Proceedings of the 2018 IEEE International Conference of Electrical and Electronic Technologies for Automotive (Automotive 2018), pp. 1–6 (2018). https://doi.org/10.23919/EETA.2018.8493233
7. Cococcioni, M., et al.: A fast approximation of the hyperbolic tangent when using posit numbers and its application to deep neural networks. In: Saponara, S., De Gloria, A. (eds.) ApplePies 2019. LNEE, vol. 627, pp. 213–221. Springer, Cham (2020). https://doi.org/10.1007/978-3-030-37277-4_25
8. Cococcioni, M., et al.: Novel arithmetics in deep neural networks signal processing for autonomous driving: challenges and opportunities. IEEE Sig. Process. Mag. **38**(1), 97–110 (2021). https://doi.org/10.1109/MSP.2020.2988436
9. Cococcioni, M., et al.: Fast deep neural networks for image processing using posits and ARM scalable vector extension. J. Real-Time Image Process., 1–13 (2020). ISSN 1861-8200. https://doi.org/10.1007/s11554-020-00984-x
10. Lin, D., Talathi, S., Annapureddy, S.: Fixed point quantization of deep convolutional networks. In: Balcan, M.F., Weinberger, K.Q. (eds.) Proceedings of The 33rd International Conference on Machine Learning, vol. 48. Proceedings of Machine Learning Research, 20–22 June 2016, pp. 2849–2858. PMLR, New York (2016)
11. Chen, X., et al.: FxpNet: training deep convolutional neural network in fixed-point representation. In: International Joint Conference on Neural Networks (IJCNN 2017) (2017)
12. Shuchang, Z., et al.: DoReFa-Net:: training low bitwidth convolutional neural networks with low bitwidth gradients (2018). arXiv: 1606.06160 [cs.NE]
13. Alemdar, H., et al.: Ternary neural networks for resource-efficient AI applications. In: 2017 International Joint Conference on Neural Networks (IJCNN), pp. 2547–2554 (2017). https://doi.org/10.1109/IJCNN.2017.7966166
14. Haotong, Q., et al.: Binary neural networks: a survey. Pattern Recogn. **105**, 107281 (2020). ISSN 0031-3203

15. Russakovsky, O., et al.: ImageNet large scale visual recognition challenge. Int. J. Comput. Vis. (IJCV) **115**(3), 211–252 (2015). https://doi.org/10.1007/s11263-015-0816-y
16. McKinstry, J.L., et al.: Discovering low-precision networks close to full-precision networks for efficient embedded inference. arXiv preprint arXiv:1809.04191 (2018)
17. Su, J., et al.: Accuracy to throughput trade-offs for reduced precision neural networks on reconfigurable logic. In: Voros, N., Huebner, M., Keramidas, G., Goehringer, D., Antonopoulos, C., Diniz, P.C. (eds.) ARC 2018. LNCS, vol. 10824, pp. 29–42. Springer, Cham (2018). https://doi.org/10.1007/978-3-319-78890-6_3
18. Choquette, J., et al.: NVIDIA A100 tensor core GPU: performance and innovation. IEEE Micro **41**(2), 29–35 (2021)
19. Banbury, C.R., et al.: Benchmarking TinyML systems: challenges and direction. arXiv e-prints, art. arXiv:2003.04821 [cs.PF], March 2020
20. Le Cun, Y., Denker, J.S., Solla, S.A.: Optimal brain damage. In: Advances in Neural Information Processing Systems vol. 2, pp. 598–605. Morgan Kaufmann Publishers Inc., San Francisco (1990). 1558601007
21. Han, S., Mao, H., Dally, W.J.: Deep compression: compressing deep neural networks with pruning, trained quantization and Huffman coding (2016). arXiv: 1510.00149 [cs.CV]
22. Pau, D., et al.: Comparing industry frameworks with deeply quantized neural networks on microcontrollers. In: 2021 IEEE International Conference on Consumer Electronics (ICCE), pp. 1–6 (2021). https://doi.org/10.1109/ICCE50685.2021.9427638
23. Kalamkar, D.: A study of BFLOAT16 for deep learning training (2019). arXiv: 1905.12322 [cs.LG]
24. Wang, N., et al.: Training deep neural networks with 8-bit floating point numbers. In: Proceedings of the 32nd International Conference on Neural Information Processing Systems, NIPS 2018, pp. 7686–7695. Curran Associates Inc., Montréal (2018)
25. Köster, U., et al.: Flexpoint: an adaptive numerical format for efficient training of deep neural networks. In: Proceedings of the 31st Conference on Neural Information Processing Systems (NIPS 2017), pp. 1742–1752 (2017)
26. Oppenheim, A.: Realization of digital filters using block-floating-point arithmetic. IEEE Trans. Audio Electroacoust. **18**(2), 130–136 (1970). https://doi.org/10.1109/TAU.1970.1162085
27. Johnson, J.: Rethinking floating point for deep learning. CoRR, abs/1811.01721 (2018). http://arxiv.org/abs/1811.01721
28. Arnold, M.G., Garcia, J., Schulte, M.J.: The interval logarithmic number system. In: Proceedings of the 16th IEEE Symposium on Computer Arithmetic (ARITH 2003), pp. 253–261 (2003). https://doi.org/10.1109/ARITH.2003.1207686
29. Cococcioni, M., et al.: Fast approximations of activation functions in deep neural networks when using posit arithmetic. Sensors **20**(5) (2020). ISSN 1424-8220. https://www.mdpi.com/1424-8220/20/5/1515
30. Langroudi, H.F., et al.: Adaptive posit: parameter aware numerical format for deep learning inference on the edge. In: Proceedings of the IEEE/CVF Conference on Computer Vision and Pattern Recognition (CVPR) Workshops, June 2020
31. Ruffaldi, E.: cppPosit. https://github.com/eruffaldi/cppPosit
32. Riba, E., Nyan, P.: tinyDNN. https://github.com/tiny-dnn/tiny-dnn
33. TensorFlow (2009). https://www.tensorflow.org/
34. LeCun, Y., et al.: Gradient-based learning applied to document recognition. Proc. IEEE **86**(11), 2278–2324 (1998). https://doi.org/10.1109/5.726791

35. Tan, M., Le, Q.V.. EfficientNet: rethinking model scaling for convolutional neural networks (2020). arXiv: 1905.11946 [cs.LG]
36. Liu, W., et al.: SSD: single shot MultiBox detector. In: Leibe, B., Matas, J., Sebe, N., Welling, M. (eds.) ECCV 2016. LNCS, vol. 9905, pp. 21–37. Springer, Cham (2016). https://doi.org/10.1007/978-3-319-46448-0_2 ISSN 1611-3349
37. LeCun, Y., Cortes, C.: MNIST handwritten digit database (2010). http://yann.lecun.com/exdb/mnist/
38. Stallkamp, J., et al.: The German traffic sign recognition benchmark: a multiclass classification competition. In: Proceedings of the IEEE International Joint Conference on Neural Networks (IJCNN 2011), pp. 1453–1460 (2011)
39. Krizhevsky, A.: Learning multiple layers of features from tiny images. Technical report (2009)
40. Recht, B., et al.: Do ImageNet classifiers generalize to ImageNet? (2019). arXiv: 1902.10811 [cs.CV]
41. Everingham, M., et al.: The PASCAL visual object classes challenge 2007 (VOC2007) results. http://www.pascal-network.org/challenges/VOC/voc2007/workshop/index.html

Author Index

Printed in the United States
by Baker & Taylor Publisher Services